PRAISE FOR
THE PERFECTION POINT

"If you're interested in understanding the American character, there are plenty of texts to consult—including, of course, the Mayflower Compact, the Declaration of Independence, and Alexis de Tocqueville's meditations on this country. Here's another work to add to the mix: John Brenkus's *The Perfection Point*, about the absolute limits of human athletic performance. . . . Now we have Mr. Brenkus to help lay out ~~precisely~~ ʹhat are likely to be the 'how far.' ʹ⸱⸱⸱⸱⸱⸱⸱⸱⸱⸱⸱⸱⸱⸱ ⸱⸱⸱⸱ ⸱⸱⸱⸱ ⸱⸱⸱ he ultimate sports achievemen⸱⸱⸱⸱⸱⸱⸱⸱⸱⸱⸱⸱⸱⸱⸱ ⸱⸱⸱ guy much the way Dr. Sanjay ⸱⸱⸱⸱⸱⸱⸱⸱⸱⸱⸱⸱⸱⸱⸱ student of the little things that ⸱⸱⸱⸱⸱⸱⸱⸱ ⸱reet Journal*

"Here's a fascinating exploration of the limits of human athletic ability. The author, host of ESPN's *Sport Science,* begins with Roger Bannister's 1954 breaking of the four-minute mile. Many people had seriously wondered if this landmark could ever be achieved, but after Bannister did it, more than three hundred people followed suit over the next decade. That happened not only because Bannister proved it could be done (and so inspired others to try) but also because human beings are getting faster and stronger, and athletic techniques are constantly being refined and augmented. The book is full of startling facts: the current U.S. record for holding one's breath, for example, is a breathtaking seven minutes and twenty-one seconds. But here's the book's most arresting element: using a variety of disciplines, including physics and physiology, Brenkus extrapolates into the future, showing us when we will reach our absolute limit of performance. . . . Sure to spark debate in sporting and scientific circles, the book is engagingly written, well argued, and—even when the conclusions seem almost science fiction—entirely plausible." —*Booklist*

"John Brenkus is a truly original voice in a sea of the same. The joy of this book lies not just in reading about the science and secrets that he's unlocked but also in how it changes the way you watch sports. For years to come, we will wait and see whether people can evolve beyond these perfection points or whether science is in fact right. *The Perfection Point* will provoke and amaze you, but it will also make you wonder just how far *will* we go?"

—Scott Van Pelt, analyst for ESPN's *SportsCenter*

"This isn't just a book for athletes or fans of sports. *The Perfection Point* is for anyone who has ever dreamed of being the best at something. A groundbreaking book that throws down the gauntlet for every human, it will blow your mind and make you look at every world record in a whole new light."

—Larry Fitzgerald, All-Pro wide receiver for the Arizona Cardinals

"First John Brenkus created *Sport Science*, one of the most revolutionary programs on television today. Now in *The Perfection Point*, he continues to push the envelope by opening your mind and making you revel in the astonishing feats that humans are athletically capable of. If you liked *Freakonomics*, you'll love this book."

—Bobby Valentine, former Major League Baseball player and coach

Sport Science Predicts the Fastest Man,
the Highest Jump, and the
Limits of Athletic Performance

THE PERFECTION POINT

JOHN BRENKUS

HARPER

NEW YORK · LONDON · TORONTO · SYDNEY

HARPER

A hardcover edition of this book was published in 2010 by Harper, an imprint of HarperCollins Publishers.

THE PERFECTION POINT. Copyright © 2010 by John Brenkus. All rights reserved. Printed in the United States of America. No part of this book may be used or reproduced in any manner whatsoever without written permission except in the case of brief quotations embodied in critical articles and reviews. For information address HarperCollins Publishers, 10 East 53rd Street, New York, NY 10022.

HarperCollins books may be purchased for educational, business, or sales promotional use. For information please write: Special Markets Department, HarperCollins Publishers, 10 East 53rd Street, New York, NY 10022.

FIRST HARPER PAPERBACK PUBLISHED 2011.

Library of Congress Cataloging-in-Publication Data is available upon request.

ISBN 978-0-06-184549-9 (pbk.)

11 12 13 14 15   OV/RRD   10 9 8 7 6 5 4 3 2 1

To Lizzie, Bryce, and Arabella—the loves of my life

CONTENTS

FINDING THE LIMITS

Are there absolute limits in sports?

Is there some speed no runner will ever exceed? A home-run distance no batter will ever reach? A weight no power lifter will ever hoist above his head, even a thousand years from now?

It's a natural human instinct to wonder about limits, especially our own. At various points in your life you probably tested yourself, for no other reason than to see what you were capable of. Maybe you stood on a driving range and hit a golf ball with everything you had, not worrying too much about the direction, just the distance. Maybe you trained for a marathon, just one, to see if you could finish, and once you knew you could, you did another to see if you could do it faster.

It doesn't have to be a physical activity, either. Maybe you time yourself when doing the Sunday crossword, hoping to beat your best. Whatever it is, at some point you probably wondered how well it's possible for anyone to do it.

It's easy enough to find out, and usually pretty impressive when you do. The longest golf drive on record is a 458-yard monster by Jack Hamm at Highlands Ranch, Colorado, in 1993. The fastest marathon was turned in by the Ethiopian legend

Haile Gebrselassie in Berlin in 2008. His time over the 26.2-mile course was a blistering 2:03:59. (Stanley Newman holds the world record for the fastest completion of a *New York Times* crossword under Guinness Book conditions—2 minutes and 14 seconds—but we're going to stick with more athletic pursuits from here on.)

Impressive, to be sure, but this doesn't really answer the question we posed. World records tell us how well people have done; they don't tell us how well they *can* do. Gebrselassie's 2008 marathon is spectacular, but it isn't the best that will ever be run. We know that because every sports record gets broken sooner or later.

But where does it end? Does it ever end? Is there a limit to how fast it's possible to run the marathon or how far one can hit a golf ball?

Until 1954, many people thought that a sub-four-minute mile was impossible for human physiology. The species just wasn't engineered to do it. This was considered practically an axiom, an absolute truth, right up there with believing that the earth was flat. For years, the record hovered around 4:02.

But one day at Oxford University's Iffley Road Track, with the wind so bad he almost withdrew from the race, Roger Bannister clocked a mile in 3 minutes, 59.4 seconds. He only broke the barrier by six tenths of a second—slightly more than an eyeblink—but he changed the world of foot racing forever.

Was Sir Roger a fluke? Some sort of one-off freak of nature?

Maybe . . . for a little over a month. It took 190,000 years of human evolution to produce a runner capable of covering a mile in under four minutes. It took only forty-six more days to produce another one who outran him. In the next three years, runners went under four minutes 18 times, and within ten years, 336 people had done it.

Why? Because Bannister proved it was possible. Once athletes knew they could do it, they did it, in droves and ever faster, each new record becoming just another obstacle to overcome.

That's not a new insight. It's practically a law of nature. People have been waxing lyrical about human potential since philosophers and poets first carved their thoughts into rocks. Every time someone declares that we can go no faster, no higher, no farther, someone else comes along to prove him wrong and leaves the doubter regretting his foolishness.

The fabled Ironman triathlon consists of a 2.4-mile ocean swim, 112 miles on the bike over the lava fields of Kona, Hawaii, topped off by a full marathon of 26.2 miles. The first time that sports magazine publisher Bob Babbitt raced, he brought a sleeping bag, and he remembers picking up a newspaper from somebody's lawn the morning of the second day and reading the results of a race he was still in. Gordon Haller won with a time of 11:46:58. The following year, Tom Warren bettered that time by over half an hour, winning in 11:15:56, and talk about breaking eleven hours was the big buzz in Hawaii. Less than twenty years later, talk has turned to breaking *eight* hours; Belgian Luc van Lierde came within four minutes of doing that in 1996.

It never seems to end. The world record for the mile today is 17 seconds faster than Sir Roger's breakthrough time, and even high school kids are running it in under four minutes. Ever try to hold your breath underwater when you were a kid? Among my friends, you were pretty good if you could go 45 seconds before popping your head up, gasping, to find out your time. Bill Graham, a retired United Airlines pilot living in Kailua-Kona, Hawaii, is the current U.S. record holder in breath holding, known as static apnea. His time was 7 minutes, 21 seconds.

But there are limits. There have to be, at least until we evolve into a different species altogether. If we agree that a man will never lift a Volkswagen over his head, then logic demands that there is something lighter than a Volkswagen that he also can't lift. And something lighter than that. At some point we can't say that anymore. Do we know where that point is?

We can agree that a human will never run at 60 miles per hour on his own two feet. He won't run at 59 mph either, or 58. We're on safe ground ruling those numbers out, and it doesn't take a lot of analysis or insight to do that. But is 28 mph possible? Almost certainly. The current world record is 27, so it's not much of a leap to imagine that one more mile per hour is possible. How about 29? Probably, and we'd be shortsighted to declare that 30 is completely out of the question. But somewhere between 30 and 58, there's a point at which we can go no faster. Is it possible to figure out that "perfection point," the speed we can get closer and closer to but never exceed?

The reasons for limits differ from sport to sport. A weight-lifter's tendons are going to tear at some point even if his muscles are strong enough. The realities of muscle strength versus weight place an upper bound on the sprinter. Simple physics dictates the highest high jump or pole vault.

The reason for the limits may vary, but there will always be reasons why there are limits.

That's what this book is about: finding the ultimate limits of human performance. Not the "safe" numbers like running 60 mph or lifting a 2,300-pound car, but the perfection points we can edge closer to yet never surpass.

In other words, we're going to be as risky and perhaps as foolish as all those other naysayers who slapped limits on human performance and came to regret it. Of course, *we* think we're different. We know more than we did then, not just about science but about athletes. We're also acutely aware of how badly each new generation breaks the marks set by the old, even "sacred" records like 61 home runs in a season.

Our method is simple in concept: For each activity, we'll start with the current world record and use some science to figure out how much better it's possible to get. It's a little more difficult in practice, but have no fear: We're not going to use any jargon or fancy equations to get to the answers. If you can use e-mail or know how to store a number in your cell phone or can manage to record a television program correctly three times out of five, you'll sail through this book. (Incidentally, a three-out-of-five success rate at recording TV programs is far higher than any baseball player in history has ever had at hitting.)

We'll also give our vision for how these perfection points might be attained in the distant future, offering circumstances and conditions that lead to the best the human race can achieve. It's worth pointing out that not every one of these perfection points will be accomplished under the glare of cameras amid a media frenzy. Sport and skill have many different uses, and when it comes to perfection, sometimes the absolute best is simply what is needed to ensure survival. Ultimately, that's why the human body makes it possible.

We'll cover the fastest running speed possible, the highest basketball dunk, the heaviest weight-lift, and a whole lot more.

And if you think you have a pretty good idea of what those limits might be, you're in for some major surprises.

CHAPTER ONE

RAW SPEED

How fast can a human run?

Even if you've never seen a track meet or run a local 10K or could care less about sports altogether, give even a moment's thought to the limits of human potential and you have to wonder: How fast is it possible for someone to move over flat ground using nothing but his own two legs?

Running has ceased to be very useful in modern society.

If we want to move rapidly, we have bicycles, skateboards, Rollerblades, cars, trains, and airplanes to do most of the work for us. On those rare occasions when we do need to run on our own two feet, it's usually away from something, like a nasty dog, a rain shower, or someone trying to serve a summons, so almost all the running we're familiar with takes place in sports: baseball, football, tennis, basketball, soccer. In all of those, there are advantages to being able to run well, like getting to the ball, basket, or goal faster than the other guy. Running is a means to an end.

Running just to run is a different story. As little of it as most of us actually do, running for its own sake still seems to hold a certain fascination, probably because it's an elemental combination of strength and speed that echoes what is likely the oldest form of competition known to humankind: "Hey! Beat you to the corner!" Nobody had to teach your five-year-old to shout that to his buddy as they headed home from kindergarten. It's basic, requires no equipment, ends quickly and definitively, and the rules are as simple as rules can get: We're both *here*; whoever gets *there* first wins.

Simple head-to-head running contests soon led to bunches of people competing at the same time, and before too long somebody got around to wondering who would win if everybody on the planet ran against everybody else. It would be tough to arrange a race like that, but timed races under controlled conditions do a pretty good job of answering the question to everyone's satisfaction. For any given distance—mile, 10K, marathon, whatever—we assume that the current world record holder would win if all 6.8 billion of us toed the start line at the same time.

Which brings us to the question of running's perfection point: What is the fastest possible speed that anyone could ever run?

And how would we find out? There are dozens of different competitive running events, but the most fascinating of all, and the most popular among fans and the general public, is also the simplest and the shortest. It's the 100-meter sprint, and whoever holds the world record gets the most coveted title in sports: the fastest man on earth.

It's not just hype, either. It's the literal truth, and it doesn't mean only on the day of the race. The world record holder at this distance is almost certainly the fastest human who ever lived. As of August 2008, that's Usain Bolt of Jamaica, and if he were to run the 100 meters in a neighborhood with a 25 mph speed limit, he'd get a speeding ticket.

In 1912, the world record for the 100-meter sprint was held by Don Lippincott of the United States. His time was 10.6 seconds. It would be 56 years before Jim Hines, another American, managed to do it in under ten seconds.

Bolt's current world record is 9.69.

It keeps getting faster, but it can't get faster forever. Once we've optimized the conditions, the equipment, and the athlete himself, there can't be any more improvement. Somewhere, there is a limit.

Location: Marrakesh Olympic Games

Diondre Boltano Hayes doesn't look like most Dominicans. Neither did his parents, or their parents before that, descendants of a line of athletes dating back hundreds of years across a handful of island nations, all within five hundred miles of their ancestral Jamaica. To say that the extended Hayes clan dominated short-distance track and field would be like saying the Swiss dominated watchmaking: It didn't need saying at all but was a simple fact of nature.

Even with that distinguished lineage to be measured against, Diondre Hayes was special, the culmination of a long series of genetically fortuitous marriages and a family culture that heaped attention and rewards on athletically promising progeny. The man standing at the start line of the 100-meter sprint is six feet two inches tall and weighs 192 pounds, only 4 percent of it in fat. Most of his height is in his legs; his tailor would measure his inseam at 40 inches.

It doesn't take an expert to see that Hayes is as perfectly crafted a sprinting machine as the species was ever likely to produce. His calf muscles are six inches in cross section at their thickest part and consist of 65 percent fast-twitch and 35 percent slow-twitch fibers. The ratio is slightly less (55 percent to 45 percent) for his quadriceps, the large muscles in his thighs, which are twelve inches wide. Specialized training has built up the power-generating capacity of his hip flexors to just shy of the point where they would damage his tendons.

He didn't want for the right gear, either, Hayes athletes being born with silver swooshes in their mouths. Diondre's super lightweight shoes are perfectly matched to the track surface, with carbon fiber spikes at the toe and mostly empty space in the heel. The total weight of the shoe is 87 grams. He wears aerodynamic sunglasses, no jewelry, and a skintight track suit surfaced with thousands of tiny dimples. His head is shaved bald.

On race day at the Marrakesh Games, there is a breeze at his back of two meters per second, about 4.4 mph, the maximum allowable under Olympic standards. The track in the hills above the Moroccan city is at an altitude of 3,280 feet, and the straight stretch used for the 100-meter sprint faces west; when the race begins at 8:00 a.m. the morning sun will be low in the sky behind him, not in his eyes. The barometric pressure is 29.14 inches of mercury, the humidity is a desert-dry 11 percent, and the temperature is 82 degrees Fahrenheit. Because of his fast times in the semis, Hayes has been assigned to Lane Four for today's final. The other lanes are occupied by runners who look a good deal like him and are capable of running the same speed.

Directed to take his place, Hayes kneels and plants his feet against the starting blocks, leans forward, and positions his fingertips on the track surface just behind the start line. Then he lets his knees settle on the track and wills himself to stillness. When the starter calls "Ready!" he comes off his knees, raises his hips high in the air, and leans forward on his fingertips.

His reaction time to the gun is perfect. The sensors built into the starting blocks detect an increase in pressure exactly a tenth of a second after the gun goes off. Just under 1.4 seconds later he's 10 meters down the track, moving at 9 mph and still hunched over, still accelerating.

At 20 meters, he sees the other runners in his peripheral vision. They're all ahead of him.

Concerned, Hayes grits his teeth and concentrates, making sure to maintain his perfect running form even as he slams the pedal to the floor and begins to assume a more upright position. By the time he gets just past the halfway point of the race he's fully upright, covering more than eight feet of ground with every step and traveling at a blistering 29.4 mph. That's the top speed of which he's capable. He isn't going to get any faster, and now that his speed has stabilized, he risks devoting some small portion of his concentration to checking his peripheral vision again.

He's still in last place.

Hayes doesn't panic, doesn't waste precious mental energy trying to make up the deficit between himself and the other runners. He has a plan, he knows his competitors, and he knows that they can't get any faster, either. But while he knows all of that, his challenge now is to make himself believe it. He has to bank on the fact that he can hang on to his high speed for a longer time than the others can hang on to theirs. If he does that, he can win.

At the 80-meter mark he's still in last place. But he's closer to the leader now, and gaining on him. Less than eight seconds have elapsed since the start of the race. For the first time he feels a strain in his shoulders as his arms pump madly to counterbalance the shift from one foot to the other and back again, almost five times per second. He also feels himself slowing down, and pours every last ounce of energy he has into his legs, not to regain higher speed but to minimize the rate at which he's decelerating. At 90 meters he's neck and neck with the front-runners. They know he's there, and he knows they know, and he also knows that they're every bit as tough, determined, and talented as he is.

At that point, his mind pulls completely away from them. The white strip indicating the finish line is the only thing in his line of vision. He has nothing at all left in the tank, but somewhere deep down in his gut he finds a drop of gas he never knew was there and seizes on it, forcing it into his afterburner and feeling it ignite. It works; he's a hair's breadth ahead of the others and can see them react to it. But Hayes reacts, too. Energized by his lead, his adrenal glands give up their final wisp of epinephrine and wring the last measure of power out of his nearly spent muscles.

The final stride that carries him to the finish line is his thirty-ninth step of the race. It takes a high-speed camera to determine that he won, by a hundredth of a second. His time of 8.99 seconds makes this the fastest hundred ever run.

It's the fastest that will ever be run.

The year is 2909, and we've reached the limit.

On the face of it, a time of 8.99 seconds is absurd. Nobody familiar with the sport of running would believe that such a time is even remotely feasible, nor is a top speed of 29.4 mph.

It is feasible. We're going to prove it. But to do so we have to go back to the beginning and ask a question. . . .

Why does 100 meters produce the fastest speed?

Let's say you wanted to find out the fastest you could possibly run over flat ground. To do that, you'd accelerate quickly in order to pour as much energy as possible into gaining velocity and not waste it covering distance. There's not much strategy: just put the hammer down and give it everything you've got. Somewhere around the 55-meter mark, you're going to hit the top speed of which you're capable. That's what 100-meter sprinters do. For the remaining 40 or so meters, they just try to slow down as little as possible.

In a 200-meter or longer race, you'd never hit that top speed. If you did, you'd run out of gas a few seconds later and finish dead last by an embarrassingly wide margin. So the 100-meter record holder is rightly considered to be the fastest man on earth because that race mimics exactly what you'd do if your goal was to reach your maximum possible velocity.

Just as a presidential candidate can win the popular vote but lose the election because of too few electoral votes, it's possible to win a 100-meter race even if one of your competitors hits a faster top speed than you. That can happen if you come out of the blocks a little earlier than he did or accelerate more quickly or decelerate less at the end. If that seems odd, consider this: In a 100-meter race between the world's best human runner and Thoroughbred racehorse, the human will win easily. Even though the horse will eventually hit a higher top speed, it can't accelerate as quickly, so it doesn't reach its highest speed until the last part of the race, and by then it's too late. Add just another 30 meters, though, and the horse would leave the human in the dust.

So technically, "the fastest man on earth" isn't necessarily the one who hit the highest top speed; it's the one who covered the full 100-meter distance most quickly. But the two are so closely correlated—the runner who wins is almost always the same as the one who hits the highest top speed—that we're quite safe using the 100-meter sprint as the gold standard in exploring the perfection point of human foot-powered speed. Whatever top speed is reached in the fastest possible 100 meters is going to be the top speed reachable, period. What will that speed be?

Using purely statistical methods, some scientists have developed models that predict the theoretically fastest possible 100-meter times. While that might sound like some dubious speculation, three things make those models fairly credible.

The first is that several different researchers who arrived at their conclusions independently of one another are in very close agreement. The generally accepted conclusion is that we'll max out at 9.44 seconds, somewhere between 250 and 500 years from now.

The second thing that makes the models credible is that they've done a very good job of accurately predicting the progressive improvement of actual world-record results thus far. Being able to predict things is the hallmark of a good scientific theory, and the more far-out the prediction, the better it makes the theory look if the prediction works. By doing nothing more than sitting in his study, Albert Einstein came up with the theory of relativity, which other scientists had a hard time believing. One of the things the theory predicted was that light could be bent by gravity, which nobody believed, either. Fourteen years after the theory was published, a British scientist named Arthur Eddington took advantage of a full eclipse to see what would happen to light from a star as it passed close to the sun on its way to earth. Not only did the powerful gravity of the sun bend the light, it did it by the precise amount Einstein had predicted. Eddington sent a famous telegram to the Royal Astronomical Society that read "Relativity is right!" and Einstein became an instant celebrity.

The third reason we have for trusting the statistical models is perhaps the most compelling of all. We know the exact 10-meter split times for pretty much every major race since the advent of electronic timing. For those not officially published, we can figure them out as long as the race was captured on tape or film. Since we know the exact number of frames per second of the camera, it's a simple matter to count how many frames it took to shoot each 10-meter segment.

Once we do that, we can select the fastest splits ever run for each of those segments. For example, the fastest that the first ten meters has ever been covered, including reaction time, was 1.69 seconds by Raymond Stewart at the 1991 World Cup in Tokyo. Bruny Surin's time of one second flat for meters 10–20 at Sevilla in 1999 is the fastest ever for that segment. The best time for the third 10-meter segment is Maurice Greene's 0.89 seconds at Stockholm the same year.

What if we do that for all ten segments of the 100-meter sprint and add them up? It comes to 9.44 seconds, the exact time predicted by the statistical models as the fastest time possible.

Obviously, that doesn't mean that anyone today is capable of running that time. The splits include those of runners who were monsters off the block but didn't have high top speed, others who reached high speed but took too long to get there, and others who got going pretty good but faded quickly. All of those skew the results by giving us phenomenal splits that no single sprinter could put together in one race. It would be like trying to calculate the fastest possible marathon by adding up the best ever one-mile splits, including a 3:50 by some professional miler who wanted to be able to say he had the lead for a few minutes and then took another five hours to finish the remaining 25.2 miles.

But the 100-meter numbers are different. Each of the athletes who ran those splits was a serious competitor on the world stage who was trying to win, not a showoff like the annoying marathon "rabbits" who purposely burn themselves out staying out front for the first few miles so they can get on television. So while adding up the ten fastest splits doesn't give us a plausible estimate of what's achievable today, it does provide a good reasonableness check of the predictions for what's possible.

Add that to the other two pieces of evidence that strongly confirm the predictive accuracy of the models, and you have an impeccably logical, utterly scientific, and thoroughly validated estimate of 9.44 seconds for the ultimate 100-meter sprint.

There's just one problem. And it has a name: Usain "Lightning" Bolt.

The Jamaican running phenom must not have gotten the memo about what the researchers predicted, because his string of 100-meter sprints in 2008 not only trashed the world record, it trashed the prediction models, too. According to nearly all of them, we shouldn't have seen his Beijing Olympics time of 9.69 seconds until the year 2030.

Here's a graphic look at how badly Bolt shifted what started out as a nice, neat statistical curve. The two lowest dots below the curve in years 2007–08 are his:

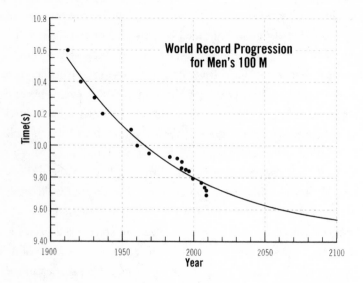

I asked University of Pennsylvania mathematician Reza Noubary, author of a textbook on sports statistics, what Bolt's Beijing performance did to the theoretical lower limit. Noubary is one of the researchers who came up with an "ultimate record" of 9.44 seconds.

"With this new data," he said with a deadpan expression, "the predicted fastest one-hundred-meter time would probably go down a little bit."

No kidding. Or, maybe the models are just plain wrong. Back in 2000, Jonas Mureika, a physicist at Loyola Marymount University in Los Angeles, approached the problem in a completely different way. He developed a statistical model using techniques drawn from seismology and accurately predicted a Bolt-like time by 2009. But nobody knew about his results because he decided not to publish his conclusions.

Why? Because he didn't believe them. "The record then was about 9.79," Mureika told me, "and my model predicted these crazy times, that by 2009 the record would be down in the high nine-sixes. I thought that was crazy. It's not going to progress that fast."

After Beijing, he must have felt the way Einstein did when the legendary theoretician stumbled onto the fact that the universe was expanding but didn't believe his own numbers. When it turned out that he was right, Einstein called it the biggest mistake of his career. "Every day that I think about that," Mureika laments in similar fashion, "I kick myself. That's my penitence for doubting the numbers."

Despite the success of Mureika's model, we have to remind ourselves that it's a purely numerical analysis and, as Mark Twain once put it, "There are lies, damned lies and statistics." Mathematical models have to be tempered with physical reality. Otherwise, you can end up with technically true but nevertheless dubious conclusions. On average, everyone in the world has one breast and one testicle, but that's not an especially useful piece of information.

Peter Weyand, a physiologist at Southern Methodist University in Dallas who is an expert on the biomechanics of running, believes that mathematical models such as Mureika's can't really predict how fast humans might eventually run.

"Predictions like that are fun," he told me, "but it's not a scientifically valid approach, because they assume that everything that has happened in the past will continue in the future." That doesn't work in the stock market and it doesn't work in sports, either. "Mathematical models can't predict what's going to happen at the extreme edges of athletic performance," Weyand observed, "the freaky phenoms that surprise us all."

Weyand is right. The statistical approach isn't going to work for us. We need a more direct method of answering our question.

In the fictional description of the fastest possible 100 meters at the beginning of this chapter, our runner took 8.99 seconds to complete the distance. Let's take a look at how we got there.

The most scientifically credible way to arrive at a realistic prediction of the fastest hundred that could ever be run is to start with the fastest that's already *been* run and then apply what we know about physiology, physics, and the lore of the track to figure out how much faster it's physically possible to get.

But we're not interested in assembling a Frankenstein with the legs of a weightlifter, the height of an NBA center, and the upper body of an anorectic ballerina. Our goal is to imagine a perfect yet perfectly plausible sprint specialist who has all the benefits of genetics, nutrition, training, mental discipline, and the knowledge acquired by his predecessors.

For a race so simple—start at point A and get to point B as fast as you can—the hundred is surprisingly complex. It consists of four distinct phases:

1. Reacting to the gun
2. Getting out of the blocks
3. Accelerating to top speed
4. Hanging on for dear life at the end

To predict the ultimate time, we'll use Usain Bolt's record-breaking 100-meter sprint at the 2008 Beijing Olympics as our starting point, then break it down into these phases. For each phase, we'll look at how much better Bolt would have done if the conditions had been more favorable. That will give us a new start-

ing point, one that still doesn't take into account the athlete himself getting any better.

Then we'll consider how much faster it's possible for a sprinter
to be. If we factor that in with the ideal race conditions we came
up with in determining our starting point, we will have arrived at
the fastest 100-meter sprint it's possible for a human to run.

We're going to assume that the sprint takes place under race
conditions. Without direct and visible competition, times would
be far slower. Send Bolt down a track solo and he wouldn't have
a prayer of coming anywhere close to a record time. As a matter
of fact, he wouldn't even have a respectable time. At those speeds,
there's simply no way to gauge how fast you're going other than in
comparison to what your competitors are doing. It's why the center lanes are so coveted and are always assigned to those with the
fastest preliminary heats. They give the athlete the best view of as
many of the other runners as possible. If Bolt hadn't had Richard
Thompson of Trinidad and Tobago pushing him from Lane Four,
it's not likely he would have run as well as he did.

Phase 1: Reacting to the Gun

The first order of business for the 100-meter runner is to start running as soon as the timing clock does.

In the old days, the man with the starting pistol stood off to
the side and fired away, which was simple enough. But with races
being won or lost by as little as one or two hundredths of a second,
it's not that simple anymore. If the starter is standing in the infield
ten feet from Lane One, he's forty-four feet away from the runner in Lane Nine. Sound travels about a thousand feet per second
in air at sea level, so when the gun goes off, the sound will reach
Lane Nine .029 seconds after Lane One, which may not sound

like much, but it would have meant the difference between second and third in Beijing. (Incidentally, if you sit in the stands down at the finish line, the sound of the gun won't reach you until nearly a third of a second after it goes off. By the time you hear it, the runners will already have left the blocks. That's why hand-timers in the old days were trained to start their stopwatches when they saw a puff of smoke from the pistol.)

One solution might be to place the starter in the middle lane far behind the runners, but that's not so simple, either. The gun also starts the clock, and there would have to be a calculation of how long the sound took to get to the racers, which would involve factoring in the temperature, humidity, wind speed, and barometric pressure. While such a start might be fair to the runners in terms of who won, it would make the determination of whether a world record was set somewhat problematical. And in these days of multimillion-dollar endorsements, it wouldn't just be an academic issue, either. In 2001, pro golfer Casey Martin went all the way to the Supreme Court over whether he should be allowed to ride a cart in tournaments because of his bad leg, so it's not much of a leap to imagine a runner filing suit over whether sound waves bouncing off a television camera gave an unfair advantage to the guy in Lane Three. (Martin won his case, by the way.)

The solution that seems to satisfy everybody is having a small speaker placed behind each set of blocks that transmits the sounds of the starter's pistol to every runner at the same time. The blank round in the pistol still sets off a satisfyingly loud report for the benefit of the crowd, but that sound reaches the athletes after the ones from the speakers, so it doesn't matter.

So that's it for the reaction time issue, right? Everybody hears the gun at the same time and off they go?

Not quite. What happens if someone goes early? That's called a false start, and we need a way to detect when it happens. It used to be done by the starter, who watched the runners as he pulled the trigger and used his own judgment. Now it's done automatically, using pressure sensors built into the starting blocks. They detect exactly when the runner pushes off. If he does that before the gun goes off, it's a false start. So now are we done with this part of the race?

Not just yet. There's another problem. International rules require that each runner take off in reaction to the sound of the gun. If instead he guesses when the gun is going to be fired, kicks off based on that and happens to hit it just right, he's technically guilty of a false start. But how do we know what's in someone's mind?

We don't. What we do know, or think we know, is the shortest interval of time over which it's possible to hear the gun and push off the blocks. In international competition, that's assumed to be a tenth of a second, or 100 milliseconds (msec). React faster than that, and it's a false start. This happened to American Jon Drummond in the 2003 IAAF World Championships when he was disqualified for leaving the blocks 53 msec after the gun. (Drummond wasn't happy with the decision and didn't do much for the sport or for his fellow athletes, who were waiting for the restart, by lying down on the track and refusing to leave for nearly twenty minutes.) A short time later in a different heat, Asafa Powell was disqualified for an 86 msec start. Powell's legitimate finishing time in an earlier heat was the fastest time of the meet, so imagine his frustration when he was tossed out.

This stuff really matters. In 1991, Carl Lewis false-started during a 100-meter race. Nervous about getting disqualified if he did it again, he got off the blocks in the restart in a dismal 166 msec. Leroy Burrell's reaction time was 117 msec. Subtract the reaction times from the finishing times and it turns out Lewis covered the distance in 9.76 seconds vs. Burrell's 9.78, but Burrell won because he was faster off the blocks and reached the finish line first.

While all of this sounds somewhat complicated—and it is, if

you're actually participating in a race or trying to produce one on a world stage—determining the perfection point for this part of the hundred is very easy. Or at least it's easy if you accept the assumption that 100 msec is the fastest possible "legitimate" reaction time to the starting gun. If you do, we're done: Our perfect race starts with a reaction time of exactly 100 msec and we move on to the next phase. But can we make that assumption?

Typical elite sprinter reaction times are in the range of 120–160 msec. The fastest legitimate start on record was Jon Drummond's perfect 100 msec in Monaco in 1993. A year before that, Bruny Surin clocked 101 msec in a World Championship semifinal, which would seem to prove that Drummond's reaction time wasn't a one-off anomaly.

But if 100 msec has been demonstrated in actual race conditions, who's to say that 99 isn't possible? Or 98, or 90?

There's no solid rationale for accepting 100 msec as the absolute threshold of human ability. For one thing, 100 is an awfully arbitrary number. When you see numbers that round, it's a warning that there's not a lot of scientific precision behind them. For another, limits like that are nothing more than a challenge, just as the four-minute mile was. When elite athletes smell barriers, they go into overdrive attempting to crack them. You can bet that sometime in the not-too-distant future there is going to be a string of sub-100-msec reaction times in important 100-meter races, followed by a great hue and cry, a handful of lawsuits, and a redefinition of "false start." For all we know, there are a bunch of Drummonds and Surins already out there who are capable of consistent 90 msec reaction times but are too fearful of disqualification to push themselves. Take the psychic shackles off and what can we expect? Hard to say, although we do know that there is an absolute bottom limit at around 75 ms based on the chemistry of nerves and muscles.

But the .10-second standard is universally accepted, and that makes it very easy to determine the perfection point for this phase of the ultimate 100 meters. As test pilots are fond of saying, "You can only *tie* the record for lowest altitude." We're going to assume that our ideal athlete nails the current standard of a tenth of a second. Anything longer than that isn't perfect, and anything less would disqualify him.

Now that we know what the perfect start is and also know that it's achievable in a real-world race because it's been done (by Jon Drummond), let's return to Usain Bolt's performance in Beijing. At 0.17 (actually 0.165) seconds, his reaction time to the gun was downright pedestrian for a world-class sprinter. In fact, it was the second *slowest* among the eight men in the field. For our purposes in determining the fastest possible 100 meters, we can adjust his reaction time to the Olympic minimum of .10 seconds.

We can't always make adjustments like that. For example, we talked about the speed at which various segments of the race have been run by others. To say that Bolt could conceivably run the first 10-meter segment as fast as the fastest ever is not legitimate, because doing that would affect the rest of his race, as you know if you've ever gone out too fast in a 10K and then faded. Similar factors come into play for the other segments as well.

But reaction time is different. It doesn't take any extra effort, so it's perfectly reasonable to say that if a sprinter reacts faster, he'll finish faster. Had Bolt gotten off the line at the Olympic standard of .10 seconds instead of .17, his final time would have been 9.62. That's the first adjustment we're going to make in trying to arrive at the 100-meter perfection point.

Phases 2 and 3: Getting Out of the Blocks and Accelerating

We're going to combine these two aspects of the race, both for the sake of simplicity and because, as we'll soon see, there's not much to be gained by considering them separately.

As complicated as the rules are governing the start, once you get going, there are only two rules left: Stay in your own lane and don't take drugs.

An athlete in the 100-meter sprint has two jobs once he's out of the starting blocks. The first is to accelerate his body as quickly as he can to the highest speed he can, and the second is to slow down as little as possible once he gets to that speed. It's his muscles that provide the force that gets him going, and it's friction that tries to slow him down. Let's start with how he gets up to speed.

If you analyze the dozen or so fastest 100-meter races, you won't find any correlation at all between who covered the most distance in the first step and who won the race. After all, a fly could get out of the blocks faster than Bolt, but it's not going to beat him to the finish line. This is why we combined our consideration of the start and accelerating. While there's certainly a benefit to honing technique, it doesn't pay to go overboard on maximizing the kick out of the blocks, because a sprinter who adopts a weight lifter's training regimen—repetitive squatting with heavy weights to build up the calves and thighs—just to enhance the initial push-off risks compromising the rest of his race. He might launch himself like a human cannonball when the gun goes off, but his leg muscles will be so overdeveloped that he won't be able to move them fast enough down the stretch.

Let's turn to accelerating down the track. We'll use Usain Bolt at the Beijing Games as our example again. Except for reaction time, he's about the best example to use when talking about any aspect of the hundred. He went from 0 to 25 mph in less than four seconds. That's the same acceleration you get in a decent sports car (which might have a top speed of 150 mph).

There's a simple relationship among acceleration, force, and mass, so knowing that Bolt's mass on race day was 90 kilograms tells us how much force he was generating. But in addition to generating enough force to get his mass moving, which is the same whether he's on earth or in outer space, the earthbound athlete has the added problem of friction. A good way to understand friction is to talk about baseball.

What does a baseball commentator mean when he tells us excitedly that a pitcher just launched a 100 mph fastball? He's referring specifically to the speed recorded by a radar gun at the moment the ball leaves the pitcher's hand. That's not the same speed at which it crosses the plate, because once it's flying freely through the air, the air itself gets in the way.

Air may not seem like much of an obstacle when you're walking to the corner for the morning paper at 2 mph, but at higher speeds, it starts to feel more and more like molasses. A space shuttle returning to earth from orbit hits the upper atmosphere at 15,000 mph. Fifteen minutes later, it's down to less than 250 mph. The only thing that slowed it down was air resistance, and the friction is so immense that the shuttle would heat up and vaporize if it weren't protected by a layer of thermal tiles.

Things aren't quite so dramatic on the ground, but they still make a difference. A Nolan Ryan fastball might be moving at 100 mph when it leaves his hand, but once it's on its way, the air goes to work on it. A baseball loses about one mph for every seven feet it travels. By the time Ryan's heater reaches the plate, it's only going about 93 mph.

Air resistance and other forms of friction play a part in many athletic events involving speed, as does gravity. If a cyclist in the Tour de France didn't have these factors to worry about, he could pedal his way up to a good clip and just coast through the rest of the race. But a cyclist has to battle air and gravity as well as friction from the moving parts of his bike.

On level ground, nearly all of a cyclist's energy goes into overcoming air resistance. Two-time Olympic cyclist John Howard wondered what would happen if you could eliminate it entirely. He mounted a wind-breaking shield on the back of a race car and rode his bicycle behind it, so that he was effectively riding in zero wind. He quickly got up to such a high speed that he couldn't turn his pedals fast enough, even in his top gear. So he went home and built a special bike with enormous gears, then tried it again. Using only the power of his legs but without any air resistance to fight, he hit 152 mph. A few years later Fred Rompelberg of the Netherlands gave it a whirl and got up to 170 mph.

Air resistance affects the 100-meter sprint as well. The force with which air impedes the motion of a moving body is called aerodynamic drag, and it increases with increasing speed. Unfortunately for runners—and for race car drivers, airplane pilots, and cyclists as well—drag increases as the square of the speed. That means that when you double the speed, you quadruple the drag. A marathon runner hits a top speed of around 13 mph. A 100-meter sprinter gets up to twice that speed, and therefore experiences four times more drag from the air than the marathoner. So anything that reduces his aerodynamic drag will play a big part in the finishing time.

One way to do that is to run with a tailwind. Olympic and international standards allow for a "following" wind of up to 2 meters per second (m/sec), or about 4.4 mph. That's not exactly a hurricane: It's the speed of a brisk walk and is less than what you feel coming out of the air conditioner in your car on its lowest setting. Because the sprinter is moving much faster than that, he's never going to actually feel a tailwind, but what he will feel is less of a headwind. The impact is not trivial. Tyson Gay clocked 9.68 seconds at the Olympic trials in Oregon two months before Bolt ran 9.69 in Beijing, but he wasn't credited with the record because the following wind was 4.1 m/sec, more than twice the legal limit, like your air conditioner on medium. Without that wind, he would have run only 9.78, barely beating Maurice Greene's 1991 record of 9.79.

But when Bolt broke the world record in Beijing, there wasn't a breath of wind. That's virtually unheard of in an outdoor race. Among the twelve fastest 100-meter times ever, all but one was run with a tailwind, including a 1.7 m/sec breeze behind Bolt himself when he broke his first record in New York three months before Beijing.

Every meter per second of following wind knocks approximately five one-hundredths of a second off a 100-meter sprinter's time. Had there been a breeze in Beijing just slightly faster than the one Bolt had in New York, he might have chopped .10 seconds off his time. Add that adjustment to the one we made for his slow reaction time and we're down to a very plausible 9.52 seconds as the basis for determining the perfection point.

There is also an Olympic standard for the maximum altitude of the track, because the thinner the air, the less the resistance, as any golfer who plays on a mountain course knows. The standard is 1,000 meters (another suspiciously arbitrary round number), or 3,280 feet. Official races can be run at higher altitudes, but the results are marked with an A, which renders any records suspect. At the 1968 Games in Mexico City, which is over 7,000 feet high, world records were broken by large margins in all three sprint distances, the 100 meters, 200 meters, and 400 meters. In the long jump, another event in which aerodynamic drag is a critical factor, Bob Beamon utterly destroyed the world record with an astonishing leap of over 29 feet, a mark that wouldn't be surpassed for twenty-three years.

The track at the Beijing Games was only 146 feet above sea level. Using standard corrections for 100-meter finishing times at various altitudes, we can assume that, had Bolt been running on a track at the allowed maximum altitude instead of in Beijing, he would have taken another .06 seconds off his time. That gets us down to 9.46 seconds.

This is why the purely statistical models are almost certainly wrong. Our estimate of 9.46 seconds is not a theoretically fastest time for a human. It's just what Usain Bolt could actually have done had he reacted more quickly to the gun and raced with the strongest legal tailwind on the highest legal track. It doesn't assume that he's any stronger than he was that day in Beijing. So while we've just demonstrated that 9.46 could be done right now, the statistical models predict that it won't happen for another 200–400 years.

And we're not done yet.

Phase 4: Hanging On at the End

Cranking up to 25 mph from a standing start took Bolt less than four seconds. Squeezing out just an additional 2 mph to get him to his top speed of 27 mph took him another two seconds. Since he was probably creating roughly the same amount of force during all of that time, why did it take so long to accelerate from 25 to 27 mph?

There are a number of reasons. The first has to do with aerodynamic drag again. From 5 to 10 mph the drag on his body quadrupled, then it quadrupled again on his way to 20 mph. By the time he reached that speed, a lot of the force he was generating went just toward overcoming the air resistance at the higher pace, along with a little to overcoming friction. There's not much left after that to produce acceleration as well.

Another reason is that there are limits to how fast muscles can move. Olympic weightlifting champion Hossein Rezazadeh of

Iraq can generate enough force in his legs to stand upright from a squatting position while holding a 578-pound barbell to his chest. That's a lot more force than any runner is ever going to create, but that doesn't mean Rezazadeh is going to be competing in any sprints very soon. Aside from being far too heavy, he couldn't move his legs fast enough.

But even a sprinter can move his legs only so fast. He not only has to push off powerfully with each leg, he then needs to get that leg forward as quickly as possible to get ready for the next step. The muscles required to do that are different from the ones that supply the pushing power and, unlike in the ostrich, the fastest animal on two legs (that's not a joke—the ostrich can run *twice* as fast as a good 100-meter sprinter, or about 50 mph, and keep it up for half an hour), they're not well developed in humans. We also don't have an ostrich's muscle-tendon units that act like springs to help that rear leg move forward.

Once a sprinter reaches the limit of how fast he can take individual steps, the best he can do if he's got power in reserve is to use it to increase his stride length so he can cover more ground with each step. World-class sprinters typically take forty-three or forty-four steps over the course of 100 meters. Usain Bolt needed only forty-one in Beijing, covering well over eight feet with every step in the middle of the race. That's about the distance from the floor to the ceiling of an average-size room.

Bolt hit his top speed of about 44 m/sec, or 27 mph, somewhere after the 50-meter mark, which is where most sprinters reach their top speed. About 10 or 20 meters later, they begin slowing down. While it's difficult to tell exactly how fast they're moving at any point in the race, we know their splits—the time they took to complete each 10-meter segment—with great accuracy. Here are the splits for seven record-breaking 100-meter sprints (reaction time is included in the 0–10-meter segment):

	Ben '88	Carl '88	Mo '99	Mo '01	Tim '02	Asafa '05	BOLT '08
RT	0.132	0.136	0.162	0.132	0.104	0.150	0.165
0-10m	1.83	1.89	1.86	1.83	1.89	1.89	1.85
10-20m	1.04	1.07	1.03	1.00	1.03	1.02	1.02
20-30m	0.93	0.94	0.92	0.92	0.91	0.92	0.91
30-40m	0.86	0.89	0.88	0.89	0.87	0.86	0.87
40-50m	0.84	0.86	0.88	0.86	0.84	0.85	0.85
50-60m	0.83	0.83	0.83	0.83	0.83	0.85	0.82
60-70m	0.84	0.85	0.83	0.83	0.84	0.84	0.82
70-80m	0.85	0.85	0.86	0.86	0.84	0.84	0.82
80-90m	0.87	0.86	0.85	0.89	0.85	0.85	0.83
90-100m	0.90	0.88	0.85	0.91	0.88	0.85	0.90
TIME	9.79	9.92	9.79	9.82	9.78	9.77	9.69
			Courtesy of SpeedEndurance.com				

Two things pop out right away. The first is that Bolt had the fastest 10-meter split in history, at 0.82 seconds. (This despite the fact that Ben Johnson and Tim Montgomery were disqualified for using performance-enhancing drugs.) The second is that he had *three* of those in a row. What shocked the experts more than anything else was not how fast he was moving, but for how long he was able to hold on to his top speed. As the next graph shows, for a 50-meter stretch starting at the 40-meter mark, Bolt never dropped below 42 km/hr, or 26 mph.

When Bolt hit his top speed just past halfway through the race, one of two things was going on: He was either moving his legs as fast as he could and therefore additional power wasn't going to do him any good, or he was putting out as much power as he could possibly generate and it wasn't enough to make his legs move faster. Where would a future athlete go from there?

To find out, I consulted with sports medicine specialist Dr. Bassil Aish, chief medical advisor for *Sport Science* on ESPN. For-

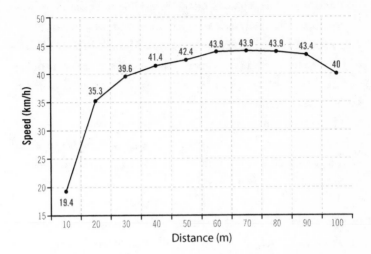

mer team physician for UCLA football, track and field, soccer, and swimming, Dr. Aish is one of the world's foremost researchers and authorities on the science of running. I asked him to speculate about the ultimate sprinter and explain his thinking.

While there is a relatively straightforward relationship between muscle mass and strength, assuming ideal oxygen utilization, the situation gets vastly more complex when you also consider speed, weight, and endurance, and even more convoluted when you get into fast-twitch versus slow-twitch fibers. Make the muscle bigger and it gets stronger, which is good for a weightlifter, but in a runner there's a point where it becomes so heavy or unwieldy that it becomes less flexible and decreases the speed of contraction, thereby actually slowing him down. Get the wrong mix of fiber types and the marathoner might be happy, but the sprinter could be a dud.

Aish has calculated the ideal muscle mass and fiber ratio for a marathon runner. We asked him to do it for a 100-meter sprinter. It turned out to be a harder task than he first thought, because of important differences in muscular demand between the acceleration and high-speed phases of the race. The explanation of how he conducted his analysis is very technical and difficult to convert into a specific speed prediction, so I asked him to express his conclusions about the "ideal sprinter" via comparison to Usain Bolt.

"He's fairly close," Aish reported back, "but there's plenty of room for improvement."

How much room, exactly? "A sprinter who achieved ideal proportions," he says, "could accelerate to a high enough speed and hold it for a long enough time to run the 100-meter distance 3.7 percent faster than Bolt could."

Aish stressed that this didn't mean that the overall result of the race would be 3.7 percent faster than Bolt's Beijing time, because that would depend on the track conditions, wind, elevation, and reaction time getting out of the blocks. His conclusion was only that, once under way, under identical conditions, the ideal sprinter will cover the distance 3.7 percent faster than Bolt.

It's a dramatic conclusion, but it comes with a caveat.

"We're talking about something that might happen a thousand years from now," Aish warned. "That's thirty generations. At the rate that humans are evolving, we're not the same species we were even two hundred years ago, and we're definitely not going to be the same species a thousand years from now."

In other words, we may never know if he's right because our genetic makeup could change enough by then to make his calculations irrelevant. So we need to keep in mind that we're talking about the limits of humans as we know them, because there's no sure way to predict where they're going.

It would seem that we now have all the data we need to calculate the perfection point, but we're still not quite ready, because we haven't yet dealt with the single most extraordinary—and downright bizarre—aspect of Bolt's Olympic race:

He wasn't even trying his best.

We know that because of what happened at the end.

Look at the previous graph and notice how much slower Bolt ran the last ten meters than the previous sixty. The reason? Eighty meters into the race, Bolt looked to the side to check out Richard Thompson, who was in second place and fading fast. Bolt, realizing that he now had the race won, dropped his arms and began celebrating, not only coasting his way to the finish and high-stepping across it, but thumping his chest for good measure.

It made for wonderful theater, and was one of the most talked-about moments of the entire Games, but what's important for our purposes is that Bolt literally quit racing with nearly 20 meters remaining. Even the most conservative experts agree that he could have gone a tenth of a second faster if he'd kept the hammer down instead of showboating.

Hans Eriksen and his colleagues at the Institute of Theoretical Astrophysics at the University of Oslo went even further. They considered factors such as Bolt's position, acceleration, and velocity in comparison with second-place finisher Thompson. Their conclusion? Had the Jammin' Jamaican not slowed down to start celebrating before the finish, he could have gone 9.55.

Much as we're tempted to use that lower figure, we won't, precisely *because* Eriksen took into account Bolt's position relative to Thompson's. Had Bolt reacted to the gun faster, he wouldn't have been as far behind in the early part of the race and therefore might not have been as motivated to pick up the pace. So it would be unscientific to simply take Eriksen's analysis at face value and make an adjustment of .15 seconds. On the other hand, it does give us firm grounds for accepting the .10-second estimate of what Bolt's exuberant display cost him in the end.

Now we're ready to do the final calculation.

We start with Bolt's actual finish time of 9.69 seconds.

From that we subtract .07 seconds to account for his poor start and get him down to the Olympic reaction time standard.

Next we subtract .10 seconds because that's the advantage he would have had with a legal wind at his back.

The track was far lower than the allowable standard of 3,280 feet. That cost him another .06 seconds, which we can subtract.

Then we take off the .10 seconds that his celebratory showboating added to his time.

We now know what Bolt's time would have been had all the conditions been ideal: 9.36 seconds.

Let's pause for a reality check, because those of you familiar with the sport of running are probably already shaking your heads. Listen to NBC commentator Ato Boldon, winner of four Olympic medals in the 100-meter and 200-meter sprint and one of the keenest observers of the sport: "You put the wind he had in New York behind the 9.69 here," Boldon said the day after the race, "and we could be down in the 9.5s, except that he shut down with 20 meters to go. So now, I'm like, is that in the 9.4s? It's mind-boggling."

There's more. Asafa Powell was twenty-five when he set the world record that Bolt broke. (He actually broke it twice; the first time was three months before the Olympics, in New York, in a time of 9.72 seconds.) Maurice Green was twenty-five when he set his record; Leroy Burrell, twenty-seven; Donovan Bailey, twenty-nine; Carl Lewis, thirty.

When Bolt stunned the world in Beijing, he was only twenty-one. And in case I forgot to mention it, the 100 meters isn't even Bolt's specialty. The 200 meters is. He didn't start running the hundred until the year before the Beijing Olympics, and when he set that world mark of 9.72 in New York, making him the youngest record holder ever, it was only the fifth time he'd raced that distance. To say that he's got plenty of time and plenty of potential to get even faster is an understatement. All of which makes the 9.36-second adjusted time seem a little less impossible.

And we're not even going to talk about the fact that one of Bolt's shoes was undone when he ran in Beijing.

With 9.36 as our starting point, all that's left is to factor in how much faster than Bolt a future athlete might be. Based on what we know about muscle weight, strength, and mobility, Professor Aish told us that the greatest possible physiological improvement we can expect is 3.7 percent.

Apply that to our adjusted starting point of 9.36 seconds, and it brings us to 9.01 seconds.

And now, at last, it seems that we're done. We've run out of reasons to think a 100-meter race could get any faster. There are simply no more sources of additional speed.

Except one. And this is where we're going to leave the safe arena of science and venture into the netherworld of pure speculation. Because there's something special about that predicted time of 9.01 seconds that's impossible to ignore.

In the middle of the nineteenth century, two men working for Sir George Everest set out to measure the height of the world's tallest mountain. Their result after nine solid years of incredibly precise and painstaking work? 29,000 feet. *Exactly* 29,000 feet. Sir George knew they were right but was so afraid that people would assume that such a round number was just a guess that he arbitrarily slapped on another two feet before he published it. Mt. Everest's official elevation of 29,002 feet stood for nearly a hundred years.

Round numbers might invite suspicion, but they're also irresistibly tantalizing. The world wasn't mesmerized by the quest for a 4:01 mile or a 3:59. It was the "Four-Minute Mile." A 99.8 mph fastball is an extremely impressive pitch, but it doesn't get a roar from the crowd. An even hundred does. When Wilt Chamberlain broke the all-time record for points by one player in a basketball game in 1962, there was still time left on the clock, so his teammates did everything they could to help him get to 100 points. That's the mark everyone remembers, and hardly anyone can recall what the previous record was.

At some point in the future, athletes are going to begin edging very close to the time we so carefully derived for the 100-meter perfection point: 9.01 seconds. When they do, a nearby round number is going to raise a very big question in the minds of athletes and track-and-field fans the world over: Is it possible to break nine seconds?

Science says no. Nine seconds isn't possible, because 9.01 is the calculated perfection point and how can you get better than perfect?

I don't know. But we can. To get from 9.02 seconds to sub-9 in the hundred, a .02 percent improvement in performance is needed, a fifth of one percent. It's actually a pretty large figure at the thresholds we're talking about, but the history of human breakthroughs tells us it doesn't matter. If you look at other barriers that were previously labeled unbreakable and then examine how far we went beyond them, you'll see that being less than one percent away from anything in sports practically guarantees that we're going to get there eventually.

As late as 1999, no woman had ever broken 2:20 in the marathon. The world record stood at 2:20:43. Less than four years later, Paula Radcliffe ran the London Marathon in 2:15:25, an improvement of 4 percent over that seemingly impenetrable barrier. The difference between the four-minute mile that Roger Bannister cracked in 1954 and the current world record represents an improvement of 7 percent. In the mid-1950s, going under ten seconds in the 100-meter sprint was considered impossible, but Usain Bolt's Beijing time of 9.69 is more than 3 percent better than what turned out to be a psychological rather than a physical brick wall. So is going from 9.01 to 8.99 in the 100 meters, an improvement of only .02 percent, really that hard to believe?

It isn't, but where could it come from? Maybe from a muscle spasm, or a midrace gust of wind that doesn't register on the official meters, or an equipment improvement. Maybe our improvement factor of 3.7 percent is wrong. Maybe it will come from nothing more than population growth: Right now there are about 6.8 billion people in the world, but in fifty years, there will be 10 billion, giving us a 50 percent larger pool to draw our best runners from. Maybe it will come from new technology: NASA reportedly spent more than $25 billion over the eight years it took to get us to the moon, so surely $25 billion in R&D over the next eight *hundred* years by the likes of Nike and Reebok can shave another .02 seconds off the 100 meters.

I don't know where we'll find that last three hundredths of a second, but what I do know is this: Come that close to a number like 9.0, and betting against it will be like betting against the sun rising in the morning. By 1954 the record for the mile had hovered around 4:02 for many years. Had it been 4:13, we might not have seen four minutes broken in our lifetimes. If our calculated perfection point for the 100 meters had come to 9.23 seconds or 9.17 or even 9.12, I would have left it at that and bet everything I had on the science.

But at 9.01 seconds, it's simply not an option for the species to leave it lying there. If we ever get to 9.01, we're going to break 9.0. We can't say when it will happen—the perfection point isn't about a detailed schedule of predicted world records, just the absolute limit—but it will, because our desire to overcome seemingly insurmountable obstacles may not be quantifiable, but it's still a bigger factor than any of the others we've considered.

There's an obvious logical fallacy in arguing that something is possible based only on the fact that we thought it wasn't. "They laughed at Columbus" is a frequently invoked rationale for some pretty dumb notions. A human is never going to run as fast as an ostrich or high-jump over the goal posts at the Superdome or lift a Volkswagen over his head. But that's not the argument here, because nobody has ever come close to doing those things.

The argument is that, at least in sports, how close we are to a barrier is sometimes all we need to know about whether it's possible to break it. Ten percent away, it's anyone's guess. Less than one percent and it's a virtual certainty, for no other reason than it's in our nature to find a way, as Bannister, Radcliffe, Sir Edmund Hillary, Charles Lindbergh, Roger Maris, and countless others did.

When a gauntlet like that is thrown before the human race, it's impossible for us to resist, and it's inconceivable that we'll fail.

The perfection point for the 100-meter sprint is 8.99 seconds. For a fraction of a second at the 55-meter mark, the athlete will be moving at 29.4 mph.

Unless the species itself changes, it's the fastest a human will ever run.

Update: On August 16, 2009, Usain Bolt ran the 100 meters in Berlin in a much-anticipated race that included American record holder Tyson Gay. This time there was no slowing down and no showboating. In the same stadium that saw Jesse Owens embarrass Adolf Hitler in 1936, Bolt ran his heart out all the way to the finish and utterly destroyed the record he'd set the year before in Beijing, leaving the track world in shock and reporters nearly speechless. His time was 9.58 seconds, an incomprehensible .11 faster than his Olympic performance. ESPN track and field expert Larry Rawson's breathless comment: "It's like he's thirty years before his time."

Actually, he's pretty much right where we predicted he should be.

CHAPTER TWO

PUMPING IRON

The limits of pure strength

There are dozens of weightlifting exercises, from the squat to the curl, but when two men meet and discover that they both lift weights, the first question asked is always the same: "So what do you bench?" The bench press is the standard of brute strength and, based on gains that continue to be made, there seems to be no limit to how much tonnage a man can get off his chest. But, of course, there is. . . .

Selenginsk, Russia
September 17, 2344

   As late as two centuries into the third millennium, humankind's ability to predict earthquakes still lagged well behind its ability to predict a coin toss. In the poorer cities and towns of the world, there wasn't much to be done about it anyway. When you're struggling to feed your family, complying with construction standards isn't the highest priority.

That's how it was in the dying town of Selenginsk, a forgotten stop on the Trans-Siberian Railway. Beset in turn by fires, floods, and Mongol invaders, the beleaguered inhabitants had all they could do to keep their houses standing, so the massive earthquake that struck the region just before dawn on a mild September morning leveled over half its structures. That was apparent as soon as the shock passed and the surviving townspeople began stepping out into the dusty streets. What they didn't know yet was that what meager emergency personnel and equipment might have been available would be diverted to the regional capital of Ulan Ude forty miles to the east. They were on their own.

At the east end of Pushkin Street, Vasiliev Podgorin tried to claw his way through a dozen people blocking the entrance to the Church of the Three Saints, or at least where the entrance used to be. The ancient one-story structure was of wood and stone construction and had been sturdy once but, having been moved from its original location to protect it from the endless series of floods, its structural integrity had been badly compromised. It lay in ruins.

The people blocking Vasiliev Podgorin's way were doing it on purpose. His nine-year-old daughter Irina had entered the church less than ten minutes before the quake. A passerby on his way to help put out a kerosene-fueled fire in the town clerk's office had heard her piteous cries. He'd called out for help and for the past hour he and several neighbors had dug frantically at the rubble, eventually uncovering Podgorin's badly frightened daughter, who was lying on her side and for some reason seemed unable to move toward them. Someone had gone to fetch Podgorin but by the time he'd reached the church, the men trying to free his daughter had made a terrible discovery: The young girl's arm was trapped underneath a massive crossbeam above her head. The beam itself was supported by other, smaller beams that had fallen along with it. The scant three feet of vertical space between the ground and the main beam was in imminent danger of disappearing altogether at the slightest disturbance.

Podgorin was screaming at the people holding him back, demanding to know why they'd stopped digging, refusing to accept their explanation that, short of amputating the little girl's arm, there wasn't anything they could do. Even if they were somehow able to squeeze cutting tools into the small space and figure out how to maneuver them so as to saw through the wood, it would do no good: The wood was the only thing keeping the rubble above from collapsing the tiny chamber altogether.

The only conceivable way to free Irina was by raising the main crossbeam a few inches so her arm could be withdrawn. Two men were in there now trying to do just that. They'd tried it with their hands, then with crude levers fashioned hastily of crowbars and bricks, but there wasn't enough room to set them up. After half an hour of trying, they managed only enough movement of the beam to dislodge a thin trickle of dust. They were out of ideas.

"Shore it up!" Podgorin shouted as he freed himself from his well-intentioned neighbors. "Shore up the timbers so they don't fall, then we dig from above!"

It was a good idea: If they could put together some vertical supports, it would diminish the likelihood of the small space collapsing in on itself, buying them some time to carefully remove the weighty debris from the outside and dig down to the girl.

At the exact moment that Podgorin leaped toward the church entrance to lead the effort, an aftershock rumbled beneath their feet. The two men who'd been trying to lever the crossbeam above Podgorin's daughter flew out of the small hole leading into the entrance tunnel as creaks and groans from straining wood combined with the rattle of loose stones to portend a horrible end to the crisis.

Before anyone could stop him, Podgorin got down on his knees and crawled into the hole, arriving at his daughter's side as the last tremblings of the aftershock died away. Pausing only long enough to take the girl's free hand in his own and kiss her forehead, he looked around to assess the situation, and saw in a few seconds that the others had been right: The only hope for Irina was to raise the heavy crossbeam, but he couldn't see a way to do it. Even if enough men were willing to crawl into the hole and team up to use muscle power, there was barely enough room for one man to position himself under the beam, let alone the four or five that would likely be required.

His daughter was saying something to him, but he ignored it and tried to think. When she persisted, he leaned in close to listen.

"Cirk chelnyek," she rasped in a voice hoarse with dust and fear.

Podgorin's eyes threatened to water at his daughter's plea for a "church man." He told her gently that he hadn't seen the priest anywhere around, and wondered if the aging cleric himself was buried elsewhere under the church.

Irina shook her head and repeated herself, forcing her voice to greater clarity. *"Cirk chelnyek!"*

Podgorin was puzzled. Circus man? Was his daughter losing her grip on reality as a result of the extreme stress?

One look at her face told him she wasn't, and in another moment he understood. Nodding at her, he turned and crawled quickly toward the opening to the outside. When only his head had emerged, he yelled, "Andreyovitch!" to the onlookers. "Go get Yevgeny Andreyovitch!"

Far from a "circus man," Andreyovitch was the dying town's only legitimate claim to fame, a power lifter revered as the greatest bench presser who'd ever lived. Because his sport was not an Olympic event, and because its practitioners couldn't enter any competitions in which performance-enhancing drugs were banned, Andreyovitch was famed primarily within the relatively small confines of elite power lifters, but within that space his name was whispered with hushed reverence in gyms all over the world. What endeared him to Selenginsk more than anything was his desire to spend as much time there as his busy schedule would allow. And Podgorin knew he was in town right now.

They found him singlehandedly pulling away fallen stones that had blocked access to the town's single-pumper volunteer fire station, and explained the situation to him as they ran back to the church. The crowd drew back at his approach, not just to give him room but in amazement. They'd all known him for years, but at five feet ten inches tall and 378 pounds, the sight of him was still startling. His upper arms were thicker than most men's thighs and the circumference of his massive chest was six inches more than his height. The immediate concern was whether he'd even be able to fit into the small opening leading to Podgorin's daughter.

Heedless of rocks scratching at his chest and back, Andreyovitch made his way inside, answering the trapped girl's smile with one of his own. Then he took a look around.

"Can you lift it?" Irina asked.

There was no point in being dishonest to one of the town's brightest children. "I have no way to know." *But I'll try*, there was no need for him to add.

He looked at her arm caught beneath the beam. "It's going to hurt."

"It hurts now," Irina shot back.

As Andreyovitch carefully rolled himself onto his back, Irina reached to the side and tried to grab a loose piece of wood. Seeing the movement, Andreyovitch nudged it closer to her with his foot. "What are you going to do with that?" he asked. "Bite down on it?"

Irina rolled her eyes as she stood the wood upright. "Don't be silly. I'm going to put it under if you move it just a little."

Andreyovitch caught her meaning. If he could budge the crossbeam, Irina might be able to jam the wood in to hold it in its higher position while he got ready for a second try. "Good—"

The words died in his throat as they felt another aftershock. They were running out of time.

"You're going to hear me let out a terrible roar," Andreyovitch told her. "That's when I'll be pushing." He put his hands up and laid his palms at the edge of the giant beam, whose lower edge was only inches from his chest. "If I move it even a little, you have to try to pull your arm out. Understand?"

Irina nodded.

"Even if it hurts," Andreyovitch added somberly. "You'll only have a second; two at the most."

"OK." Irina shifted her weight to buy some purchase with her hip. Then she dropped the chunk of wood she was holding; there wasn't going to be any second chance.

Andreyovitch took his hands down and spit on them, then placed them back on the beam and looked over at Irina. "Ready?"

"Yes." Any trace of bravery was gone from her voice, but her resolve was firm.

The strongman looked back up and took a deep breath, then let it out slowly. He did it again, and again. Then he took the deepest breath of all, held it for a second and let it out in a guttural, bloodcurdling scream as he pushed upward with all his might.

As the raw power flowing through his muscles was focused and channeled into a massive upward thrust, his efforts were answered by a series of cracking, groaning sounds that cascaded from the main beam through other wooden supports that had been connected to it for centuries. His eyes squeezed tightly shut from the superhuman effort, Andreyovitch had no way to know whether Irina had freed herself but it made no difference: Almost as soon as he'd straightened his arms after pushing the beam several inches higher, he felt like his bones were about to snap, so he lowered his hands, braking the downward motion of the beam as much as he could, praying that the tons of rubble just above his head wouldn't come crashing down on him and Podgorin's little girl.

They didn't. But despite saving Irina Podgorin's life, Andreyovitch would never get the credit he deserved for the greatest bench press in history, a feat that could never be surpassed.

When I was a teenager, athletic bragging rights went to the guy who could do the most push-ups. We didn't stop to think about the fact that raising your body off the ground by pushing down with your hands was a lot harder for a 200-pound linebacker than a 140-pound gymnast, nor were we particularly interested in using math to calculate strength-to-weight ratios. (Come to think of it, we weren't particularly interested in using math to do *anything*. Once we hit the gym, schoolwork was the farthest thing from our minds.) All we knew was that the guy who could do fifty push-ups without a break was stronger than the guy who could only do forty.

As we got a little older, though, push-ups went out of style, especially since some of the best athletes could do hundreds of them. Let's face it: Impressing girls was the whole point, and endlessly repeating a single exercise wasn't the most dramatic way in the world to do that. It's hard to get excited about an athletic feat if you have to look at a score sheet in order to discover that something important just happened.

So at some point we turned the push-up upside down and began lying on our backs instead of our bellies and pushing barbells into the air instead of our bodies. At first it was ten reps at a time, increasing the weight week by week, but the specialists had a different idea: a single, all-or-nothing attempt to lift as much weight as you possibly could. "So what do you bench?" wasn't an inquiry into your training routine. It was essentially asking, "Just how strong are you?"

That's pretty much all the bench was good for, that and building a massive chest. Those gigantic pecs weren't actually good for anything else in sport but they sure stopped the show when you took your shirt off. At least if you didn't overdo it. World-class bench pressers, like nearly all champion weight lifters, don't look anything like Charles Atlas or Arnold Schwarzenegger did in their primes. Bodybuilders have an entirely different goal than weight lifters, and while they move some fairly massive weights around in the training room, what they're doing is sculpting muscle, not just building it. Weight lifters, on the other hand, don't care how they look; they care how much weight they can toss around. In getting there, they end up with very little of what bodybuilders call "definition," that quality that lets you see individual muscles as though you were looking at a living Renaissance statue. Weight lifters don't look like Michelangelo's David; they look like sandbags with limbs.

We often talk about the "purity" of various sports, without really defining the term. But in tackling the bench press we're going to get into one of the most elemental human physical challenges, and when it comes to brute strength, the concept of purity is worth revisiting.

There are a lot of athletic endeavors that seem simple at first glance but turn out to be remarkably complicated upon closer examination. What could be simpler than the 100-meter dash? Start here, run there as fast as you can . . . bada bing, you're done.

But as we've seen, the "simple" dash is anything but. Extended doctoral dissertations have been written about the start out of the blocks alone, and the bases for disqualification are still being endlessly debated. Leaning into the tape at the finish, a tricky and risky maneuver that takes months to master, can mean the difference between a podium finish and a sad trip back to the locker room. Wind assistance, the design of the shoes, the surface of the track, elevation, humidity—all are factors in an intricate dance that coaches and athletes spend years trying to sort out. To top it off, we're not even always sure who won at the end of the race. In close contests, photos have to be analyzed and fretted over, and controversy often accompanies the medals. There's just too much going on to call a foot race a "pure" test.

High jumping comes closer. While the technique is complex,

the event itself is simple. There is no timing involved, and deciding the outcome is easy: Either the bar stays in place or it doesn't, and as long as you don't have springs in your shoes or use a trampoline, nobody cares how you got over it, just whether you did or not.

Determining who possesses the greatest strength isn't always so easy, either. Consider the clean and jerk, a popular but harrowing event in which a contestant first has to raise a barbell to his chest, then get it up into the air over his head. The latter move is a dangerous one, in which the athlete abruptly lowers his body at the same time as he lifts the weight, then tries to stand erect to complete the move. It's a technique-intense effort in which the medal doesn't necessarily go to the contestant with the greatest raw strength.

But lifting a dead weight ought to be the "purest" test of strength because, if done right, there's no way of getting around the physics. You can't outstretch your competitor to reach a touchpad, you can't lean across the finish line, you can't strategize by backing off early in order to make a big spurt later, and there's no technique that is going to let you sneak your way to a few more pounds of pressure.

We've all seen footage of seemingly superhuman feats of strength: people ripping phone books in half or towing 707s over their shoulders or bending quarters with their bare hands. Every year on his birthday, fitness pioneer Jack LaLanne jumps into a lake and swims while pulling several dozen boats through the water behind him. These kinds of tricks don't tell us much. For one thing, the "towing" feat lets the effect of power accumulate over time. Even if you're not as fit as LaLanne, just keep swimming and eventually the boats are going to start moving. Tearing phone books and bending quarters have as much to do with leverage and technique as strength and, again, the effects accumulate over time. You can even take a rest and pick up where you left off.

Not so with the greatest of all tests of strength, the bench press.

Lie on your back, lower a barbell full of weights to your chest, pause, and return the barbell to the rack. The physics are wonderfully simple: Gravity pulls the bar down and never quits or varies. The athlete has to overcome that gravity and make the bar go the other way until it's back where it started.

There are some rules governing competition. They vary according to which organization is in control, but the basics are pretty much the same across all of them. The lifter lies on a bench with his feet flat on the floor and his head, shoulders, and buttocks in contact with the bench surface. Spotters may be used to help get the bar off the rack. The move officially begins when the spotters have let go and the lifter has sole possession of the bar with his arms straight up in the air and locked. His hands can't be spaced more than 81 cm (about 32 inches) apart, in order to limit the reduction in distance between bar and chest.

The lifter then lowers the bar to his chest. As soon as a judge determines that the bar is motionless, thereby ensuring that there is no advantage from bouncing it, he gives the audible command "Press." The lifter then has to push the bar back up until his arms are once again locked. The bar can stop during the upward portion of the lift, but at no time can either hand descend. Once the judge decides that the arms are locked, he issues the command "Rack," and the spotters are once again allowed to assist as the bar is returned to the rack.

Pretty simple, and basic physics still rules, which isn't to say that there aren't some advantages to certain body types. The shorter your arms and the bigger your chest, for example, the shorter the distance you have to push the weight to get it back up into the air. Other than that, though, the bench press is about raw, unadulterated muscle power. Of all the power-lifting events, it's the one in which competitors use the heaviest weights, even more than the squat, which uses leg rather than arm power.

But all is not quite what it seems. There is a way to sneak more power into the event, and that's by wearing a "lifting shirt." This remarkable piece of clothing acts almost like a spring, storing potential energy as the barbell is brought down to the competitor's chest, then releasing it as the bar is pushed back into the air. Lifting shirts were originally designed to prevent injury, but when it was discovered that they could actually increase an athlete's performance, the safety consideration pretty much went out the window. A single-ply shirt can increase the amount its wearer can lift by a staggering 20–30 percent. The two-ply (or more) models can provide even more but are not permitted in most competitions.

Further complicating the picture is the issue of drug testing. The differences in performance between those organizations that test and those that don't are so great as to result in two essentially unrelated sports. Add this to the shirt issue and we wind up with six different classes of competition: tested versus untested for shirtless, single-ply, and multi-ply.

With respect to the shirts, we're going to deal with the ultimate bench press in the purest terms possible, with our athlete going shirtless (called "raw" in the sport) or wearing only an ordinary T-shirt. Unfortunately, there's not a lot of data to work with. The body that governs powerlifting worldwide is the International Powerlifting Federation (IPF). The body that governs powerlifting in America and falls under the rule of the IPF is USA Powerlifting. Both of these organizations allow single-ply lifting shirts in all of their competitions, rendering those records useless. We thought of using some rules of thumb to convert shirted (or "geared") performances into raw, but it turns out that the benefit of using the shirts varies too widely, from as little as 13 percent to as high as 50 percent. The good news is that we're only concerned with record lifts, and there are plenty of raw records for us to work with.

As for the drug issue, we're going to ignore it. While we certainly don't condone the use of performance-enhancing substances in competition, what we're trying to do is determine the ultimate in human achievement, and it simply doesn't make sense to limit the means of getting there. (For a fuller explanation of how we arrived at that decision, see the chapter on performance-enhancing substances.) Both the IPF and USA Powerlifting are drug-free and do random testing as well as testing of all winners and record breakers. So while performances sanctioned by these organizations provide us with a lot of useful data, they don't directly help us get to the perfection point.

Given how much we know about the effects of steroids, it might surprise you to know that the current world records for raw (shirtless) lifts are essentially the same for tested versus untested athletes. The drug-tested record bench press of 711 pounds is held by James Henderson, and it's stood for thirteen years. The untested record is only slightly more, at 715 pounds, set by Scott Mendelson in 2005. This would seem to be a knock on the theory that steroid use has a profound impact on athletic performance.

But there are two important factors to take into consideration. The first is that we have no way to know whether Mendelson was on the juice. All we know is that his record was set during a competition that didn't provide for drug testing. While it's easy to jump to the conclusion that he therefore was "roided up," that's not entirely fair. For one thing, drug testing is expensive, and a lot of less formal, unsponsored events can't afford to undertake it, so we simply don't know whether drugs were involved.

What we do know, however, is the weight of the athletes. The reason this is important is that, in powerlifting, a key indicator of strength is how much weight an athlete can lift *relative to his body weight*. This highly respected classification is analogous to the "pound-for-pound" standing of boxers. The best lightweight fighter in the world would get pummeled by even a mediocre heavyweight, even though the lightweight is technically the better fighter. In the same way, bench press performance as a function of body weight tells us a good deal about the relative strength of lifters.

James Henderson weighs 390 pounds, so his drug-free lift of 711 pounds is 1.82 times his body weight. Scott Mendelson weighs only 314 pounds, so his 715-pound lift is an amazing 2.28 times his body weight. Without implying one way or the other whether Mendelson was on the juice, we might be tempted to say that an untested lifter with Mendelson's strength but weighing Henderson's 390 pounds could hoist a mind-blowing 882 pounds into the air.

It's not that straightforward, because there are other factors that enter it aside from a simple multiple of body weight. The specifics were worked out by Robert Wilks of the Australian Powerlifting Federation and codified as the Wilks coefficient. His formula gives the truest apples-to-apples comparison across weight classes and even genders.

Interesting, but not our concern. We're looking for the highest amount of weight that will ever be benched by a human regardless of the lifter's body weight. Where the Wilks coefficient becomes useful is in translating known performances by existing lifters into hypothetical performances by idealized athletes.

Mendelson's 715-pound record was set without a lifting shirt. Wearing one, Mendelson pushed 1,008 pounds into the air in 2006. That's a 41 percent advantage over raw. The fact that a piece of equipment can increase the amount lifted by 293 pounds tells you all you need to know about why we ruled out the use of lifting shirts in our search for the ultimate bench press. Allowing its use wouldn't be much different from allowing the athlete to use an auto jack. (The world record for "geared" lifting is 1,075 pounds, set by Ukrainian-American power lifter Ryan Kennelly. Unfortunately, we don't know what Kennelly can bench without wearing a lifting shirt.)

The greatest pound-for-pound lifting performance ever without a shirt? Undoubtedly Andrzej Stanaszek's 391 pounds in 1994. About the size of a jockey at 111 pounds, he benched 391, which is 3.5 times his body weight. That would be like Henderson benching 1,365 pounds.

Sounds insane, but it's a useful number because it gives us a starting point from which to think about the most a human could ever bench. But it's not as simple as configuring a blown-up Andrzej Stanaszek and then laying on blubber just so we can multiply body weight by some relative factor. Rules of thumb fail when you do things like that; what we need to consider are the realities of muscle power, size, and bone strength.

There's also something else worth mentioning when trying to use an existing record holder as the anchor point for analysis: The feeling in the powerlifting community is that the bench press numbers that have been achieved to date, although very impressive, are far less than they might be because many of the most talented strength athletes go into professional sports before achieving their true potential. There's not much money in powerlifting, which is an amateur sport. A good example is Mark Henry, who was an extremely successful United States Olympic weightlifter and powerlifter (a rare combination) in the 1990s who never realized his full strength potential because he cut his amateur career short to go into professional wrestling.

We also need to set some guidelines so things don't get out of hand. We mentioned before that there is a rule limiting how far apart the hands can be on the barbell in order to make sure that fully locked arms don't result in a bar that's only an inch or two above the chest (assuming someone could even lift the bar with that hand position). But even with that rule in force, think about the advantage a person with very short arms and a very huge chest has. He only has to move the bar a fraction of the distance that a more normally proportioned athlete would.

The establishment of a minimum range of motion of the bar in the bench press has been discussed by the leadership of the IPF, but no such requirement has been implemented. In the IPF and all the other federations enforcing a maximum grip width for all lifters regardless of body type, a very short person—for example four-foot three-inch Anton Kraft, world record holder in the 123-pound class—might only have a stroke of a few inches using the maximum legal grip.

The topic is not an academic one. It actually arose at this year's Open IPF Worlds. After taking her first attempt, one athlete was required to move her hands closer because the officials at the meet did not consider her lift a "true bench press" due to the lack of bar movement, even though no such requirement exists in the rules. This tells us that there will at some point be an official rule setting a minimum bar movement distance, or else someone with a weird body type is going to come along and do a perfectly legal lift of half an inch. After consulting with experts in the sport, it looks as though the standard is going to be at least six inches of vertical bar movement, so that's what we're going to use for the perfection-point lift.

The question naturally arises as to which is going to be the limiter when it comes to the ultimate bench press: the strength of the muscles to move the weight or the strength of bones and tendons to support it? In other words, is it possible to build muscles so strong that they could break bones or wreck joints by lifting weights that are too heavy for those body parts to tolerate?

It is very rare for the upper or lower arm bones to fracture in bench pressing. When it does happen, it's almost always to people using supportive bench press shirts. In a healthy person without relevant pathology, the limiting factor will almost always be muscle strength. However, in people with tendopathy, the limiting factor can be tendon strength. The primary muscles used in the bench press are the pectoralis major, triceps brachii, and anterior deltoid. It is not uncommon in heavy bench pressing for the athlete to experience tendon ruptures, a form of muscle tear in which the tendon pulls off of either the insertion (bone) or origin (musculotendinous junction) in the pectoralis major, triceps brachii, or the rotator-cuff muscles.

But the real reason for these kinds of injuries isn't that there was preexisting tendon pathology: It has to do with one of the drawbacks of steroid use that isn't exactly a "side effect" like the well-publicized ones but is more of an "unintended consequence." Steroids result in rapid muscle growth but don't affect tendons, which have to struggle to catch up "naturally" as a result of repeated use during exercise. In non–steroid users, the entire muscle-tendon-bone unit generally adapts proportionately, and musculoskeletal injuries are less common. But when steroid use increases muscle size disproportionately, tendons lag behind the growing muscles in terms of the forces they must transmit, and they become the weak link in the chain that includes both muscles and bones.

Making the situation worse is the fact that anabolic steroids exert an independent effect on the structural integrity of the tendon by causing a misalignment of collagen fibrils, thus compromising the mechanical (tensile) properties of the tendon. So not only are the tendons lagging the muscles in strength, they might actually get weaker as well. Tendon ruptures are one of the results of this out-of-balance situation and something we need to be mindful of in conjuring up our ideal bench presser.

Our ideal lifter is going to be a very big guy, at least in terms of weight. While weight lifted in relation to body weight is a highly respected competitive category, we're going for absolute poundage. (Besides, the "relative" category has some interesting facets that complicate the analysis. As an example, some of the best relative lifters are people who have had devastating diseases, such as polio, or injuries resulting in withered legs. While their jacked-up upper bodies are normally proportioned, the absence of leg weight allows them to compete in lower weight classes, making their lifts in relation to body weight far in excess of what an able-bodied athlete can achieve.)

Training focus for the perfection-point lift would be almost entirely on the musculoskeletal and neuromuscular systems. There would be little need to worry about the cardiorespiratory system because the specific task is very short and intense in nature, meaning that blood and oxygen are not limiting factors; muscle-force production and neuromuscular efficiency are. Training should center on developing larger muscles capable of producing greater force and becoming more efficient at recruiting as many muscle fibers as possible. The program would include quite a bit of training with heavy weights and low repetitions in the bench press and its variations, along with supplemental exercises for the secondary muscle groups involved (triceps, deltoids, etc.).

There are a number of approaches to estimating how much weight a perfectly constructed bench presser could theoretically lift. One is to employ comparative physiology and use animal models—for example, looking to the mountain gorilla and its strength capabilities. But humans are not gorillas, and making the leap in logic from human being to animal is a classic pitfall in comparative physiology. While looking at other mammals allows for a starting point in the theoretical process, detailed biomechanical and physiological analysis combined with mathematical modeling is required to get the right number.

We asked a number of sports scientists who specialize in strength events to undertake just such an analysis, and we were amazed at how closely their results agreed even though they worked independently.

As we'll often do in this book, we use the sport's best athlete as a starting point and go from there. Scott Mendelson at his peak weighed close to 380 pounds. He stands six feet, one inch tall, which is a disadvantage for a bench presser because the rules require full elbow extension, and so a taller lifter has to push the weight through a greater distance than a shorter athlete with the same muscle mass. The man who achieves the perfection point is going to resemble a World's Strongest Man competitor, only shorter, standing about five ten and weighing close to 400 pounds. His torso will be disproportionately long relative to his arms and legs, giving him enough length for lumbar and thoracic spine arch advantage as well as firm butt and shoulder contact with the bench. His short arm span of sixty-eight inches will ensure that he only needs to push the barbell through the minimum six inches of vertical before his elbows are fully extended.

Muscle strength is directly correlated with the cross-sectional size of the muscle, and our guy's will be huge, just like our fictional Russian hero, Yevgeny Andreyovitch. His chest and back circumference will be a massive 76.4 inches. As for his upper arms, you might think that the only thing stopping them from being tree trunks is that his bones couldn't take the strain of the weight he could push, but that's not the limiting factor. Muscle growth is constrained by a substance called myostatin, a beta protein also called growth differentiation factor 8. Myostatin is genetically encoded but there are ways around it, typically via the use of follistatin, which blocks the binding of myostatin to its receptor. Cattle called Belgian Blues that have had myostatin production genetically manipulated away have muscles 40 percent larger than normal. Occasionally there are animals and humans called "myostatin nulls" that naturally lack the genetic inhibition and consequently end up with enormous muscles.

Fortunately for our analysis, we don't need to concern ourselves with broken bones or tendon tears because there's a mechanical limit that interposes itself well before we get to that level. If upper arm circumference is greater than 33 inches, the lifter will simply not be able to move his arms through the required six inches of vertical. That upper arm circumference, by the way, is four inches more than the biggest ever recorded.

So we now have our perfection point dimensions, but it's important to note that the final lift calculation also has to assume ideal body makeup and perfect technique. With respect to body makeup, there are two factors usually taken into account when assessing athletic performance. One is the involvement of slow-twitch muscle fibers. A typical mix in most people is about 50-50, with a 5 percent swing either way. Sprinters who rely on explosive movements can be as high as 84 percent fast-twitch, which is the same as a cheetah. That high percentage of fast-twitch fibers is also essential in our bench presser. There is no endurance involved here, only a sudden burst of power that lasts a very brief time.

The other factor is aerobic capacity. Clearly, there is no aerobic component to the bench press. There's not even enough time to draw and release a full breath. What is key is technique. More muscle doesn't necessarily mean more power. A power lifter can often bench more than a bodybuilder who has more defined and bigger muscles, and it all has to do with differences in training. Power lifters have to condition their central nervous systems to recruit fast-twitch muscle fibers to do one explosive lift, and bodybuilders train to exhaust their muscles with smaller weights but multiple reps. The former builds strength; the latter builds size.

And while we're on technique, it's worth noting that the rules of bench pressing will actually result in a slight reduction in performance for our perfection point lift because of the power differential between the pectoralis major and the triceps. The pectoralis major chest muscle is thicker and wider than the triceps, with a much broader and more stable origin, including the sternum, than the triceps. If the lifter has a grip that is too narrow (his hands too close to each other), he would place relatively more emphasis on the weaker triceps muscle and would lose some lift ability. If he has a grip that is too wide, he applies more load to the pectoralis major but loses relative contributions of the deltoid. For a 5'10" lifter with a 68-inch arm span, the optimal arm placement technique would be with the index fingers placed three inches wider than the lateral aspect of the shoulder. This would be wider than the maximum allowable grip length in competition of 31.9 inches.

Assuming that the chest contributes two-thirds of the power in a lift and the arms one-third, and using Scott Mendelson as the starting point, we can use the proportionately larger chest and arm sizes of our ideal athlete to calculate the heaviest human bench press possible. I'll spare you the arithmetic, which includes a 2.3 percent downward adjustment for the grip requirement, but it comes to 304 pounds from the arms and 617 pounds from the chest.

Yevgeny Andreyovitch had no way to know it, but when he pushed upward on the wooden crossbeam trapping Irina Podgorin's arm, he'd applied a force equivalent to a bench press of a staggering 921 pounds, which is nearly the rated cargo-carrying capacity of a Ford F-150 half-ton pickup.

CHAPTER THREE

SWIMMING

How fast can someone swim the 50-meter freestyle?

We watched, riveted, as Michael Phelps cracked records left and right on his way to winning eight gold medals at the Beijing Olympics of 2008. Less noticed but equally impressive, another man set a new mark in the 50-meter freestyle event, making him the first to break the 21-second barrier and the fastest swimmer in the world. How "legitimate" was his feat, and how far was it from the fastest 50-meter free it will ever be possible to swim?

Lake Entiat, Washington
July 19, 2256

There's a technique taught to seaplane pilots called the "glassy water landing." When the water is so smooth that there's no way to tell where the surface is, you set up a slow descent and wait to touch down because you can't tell exactly when it's coming.

The swimmers now poised on the starting blocks had a similar problem. The waters of Lake Entiat were legendary among water-skiers for their mirrorlike surface, and at 6:00 A.M. on a windless August day, the reflections of the surrounding mountains were so still they might have been painted on. The temporary dock the athletes were on was supported by thin stilts so that the movements of swimmers and coaches wouldn't disturb the surface, and they'd all slept there the night before so they wouldn't need boats in the morning. The operators of the Rocky Reach Dam several miles downstream had closed the floodgates two days before to ensure that there wouldn't be even the slightest flow of water in the lake, which was technically not a lake at all but just a wide spot in the Columbia River.

Right now, Entiat was the fastest swimming pool in history. There were no walls for waves to bounce back from, and the bottom was too deep to reflect them upward toward the swimmers. The only surface disturbances they would have to contend with would be the ones they generated themselves, and those would be quickly left behind and not revisited: The race was a one-way sprint, from the dock to the laser-defined finish line fifty meters away.

There were only three swimmers in the unsanctioned race. Zheng Li Zhao, captain of an elite black-ops unit of the Chinese navy, was the current world record holder, a fanatical self-disciplinarian who demanded equivalent devotion from the men and women in his charge. He was the picture of the new militarism that had swept through China after decades of economic and political turmoil.

Jerrol Sweeney of the United States was a senior at UC San Diego, free-spirited but intensely determined, with no problem essentially carrying the ball for Western democracy, even though Jan Hellenboch of the former Netherlands felt that he was doing the same by representing United Europe. In the dangerous brinkmanship that had developed between East and West amid Third World land grabs and the latest failure of disarmament talks, the rivalry and internecine cold warfare between the United States and the UE had grown to the point where it threatened to cloud the original intent of holding China in check while averting a potentially catastrophic fifth world war.

There was one positive aspect of the three-way rivalry: Perhaps out of some innate cultural sense of self-preservation, the superpowers had focused their bitter enmity onto their athletes instead of their armies. Much as had happened during the original Cold War of the mid-twentieth century, superstars of sport had become their Kagemusha, the shadow warriors who fought so the nations didn't have to.

Zhao, Sweeney, and Hellenboch had traded the world record fifteen times in the last eighteen months. That Sweeney had gotten one of his during the New York Olympic Games ten months earlier stuck in the craws of the other two and, as though in a plot, they'd denied him the title of World's Fastest ever since. There had, of course, been no conspiracy, Zhao and Hellenboch having the same contempt for each other as their presidents had for each other. Regardless, the pressure on Sweeney had grown almost unbearable; no day went by that he wasn't on the home page of every news site in the country.

The same was true for the other two, and they were getting tired of the endless competitions and no resolution. They'd decided to settle it once and for all with one race under optimal conditions and no other competitors. All three agreed to participate in no other races during the three months preceding this one, and all three would retire afterward, with one of them crowned the best ever. They were so much faster than anyone else in the sport that no serious challenge could be mounted— ever—and the rest of the swimming community would be able to stop racing for fourth place even though no more records would be set.

The international governing body had arranged with state authorities to keep unauthorized personnel, including news media, five hundred meters from the shore of the lake and aircraft a mile away from the dock. Roads to and from the site had been jammed to a standstill for two days, and the skies in every direction outside the no-fly zone were thick with helicopters. The swimmers did their best to ignore it all but could practically feel the thousands of telescopic lenses trained on them.

The Chinese coach had anticipated the difficulty of seeing exactly where the surface of the water was and had come prepared. As the signal was given for the athletes to take their places behind the individual dive platforms, he opened a plastic jar, dipped his hand in, and came up with a fistful of flour. When he tossed it onto the water, the surface turned cloudy, and as the flour began to absorb water it congealed into tiny, sand-sized dots that provided the necessary visual perspective. The other two coaches noted with satisfaction that the flour that hung in the air also illuminated the laser that defined the point at which the swimmers would have to stop dolphin kicking and begin stroking.

"Auf die plätze!" the starter commanded softly in the German that had been agreed to during the negotiations as the official language of the event.

The competitors stepped up onto the platforms, swinging their arms, bending over, shaking their legs—last-minute routines that did little actual physical good but were part of the rituals that had served them in the past.

"Vertig!"

They put their fingertips on the front edges of the platform and knelt. When the gun went off, they exploded, trying to grab as much force and forward motion as they could from pure leg power at the only part of the race during which their legs would be the primary source of propulsion.

Once in the water, only three things mattered: generating as much power as possible, funneling it into forward motion with the least waste, and minimizing the enormous drag of the water. The last two were the product of technique, and these three swimmers had honed theirs to the point where there was no further improvement to be made. The only differences were the particular shapes of their bodies, and those differences would count: By agreement, they were wearing only the skimpiest of Speedo-type swimsuits, in order to leave no doubt that this competition was about human ability, not technology.

That left power, and it was difficult to tell who among them had worked the hardest in conditioning his muscles to deliver it. There wasn't anything more any of them could have done, so the academics who'd been following the preparations were nearing consensus on the conclusion that this race would be won by whichever of the three athletes was genetically predisposed to be the best who ever lived, or would ever live, and that whatever time he clocked today was the fastest that could ever be.

In the century that followed the conclusion of this event, no one else even came close.

What is it about water?

Without it we couldn't live, obviously, but that doesn't explain our elemental fascination with the stuff. Let a little kid loose in a plastic pool six inches deep and he'll splash around in it for hours. Put a house on a lakefront instead of a side street and its value doubles, even if nobody living in it swims, boats, or fishes. People will pay a premium just to *look* at water, or even just to know it's there whether they actually look at it or not.

And of course, as is our nature, we have a lot of ways to compete using water. We race on top of it in kayaks, shells, sailboats, powerboats, paddleboards, water skis, and hydroplanes. We jump into it from springboards and platforms, free-dive deep down into it, ride surfboards atop ocean waves, roll logs in rivers, play water polo in pools, perform synchronized ballets half in and half out of water, and even leap clean over it in the steeplechase.

At the simplest level, we swim in it, and we've been doing it for a long time. The earliest record of swimming dates back to Stone Age paintings from seven thousand years ago, and it's referenced several times in the Old Testament—for example, "They will spread out their hands in it, as a swimmer spreads out his hands to swim" (Isaiah 25:11). We swim through water in formal competitions using several different strokes and a dozen different distances in pools, lakes, rivers, and oceans. As with running, the longer the distance, the more complex the strategy, because athletes have to carefully balance how much power they generate against how long they can last at that rate. No one wants to lose a race because he held too much in reserve, but no one wants to run out of steam before the finish line, either.

There are some events where none of that matters. They're so short that the only strategy is to swim like a fleeing suspect, generating as much speed as possible from the start to the finish because it's over too quickly to worry about conserving energy. The premier, damn-the-torpedoes glamour event when it comes to that kind of raw power and speed is the 50-meter freestyle. In the middle of an international-class 50-meter free when the competitor hits his top speed, he's swimming as fast as it's possible for him to move through water using only his arms and legs.

There are two versions of the 50-meter free. The one called the long course is done in a fifty-meter pool. The swimmer dives in at one end and swims a straight line to the other. The short-course version takes place in a twenty-five-meter pool. The swimmer dives in and swims to the opposite end, where he turns, kicks off the wall, and heads back to where he started. We're going to talk only about the long course, because the turn and push-off of the short course makes the long course the purer of the two in terms of raw swimming speed.

When we determine the perfection point of the 50-meter free, we'll also be finding out the fastest a human will ever be able to propel himself through water.

••••

When swim coaches analyze races in order to improve a swimmer's performance, they generally divide them into several parts: the start, pure swimming, the turn, and the finish. There's no turn in the 50-meter free, so we'll just look at the other three.

Racers enter the water by diving off a small platform. This is by far the fastest part of the race, because the body moves through the air, where there is almost no resistance. For analytical purposes, the start is considered to be the first 15 meters of the race. After that, the speed that resulted from the dive runs out, and pure swimming takes over. The effectiveness of the dive is a function of strength and technique. The more powerful the push-off from the platform and the better his technique once he hits the water, the farther the racer will travel before he has to start coping with the enormous drag of the water by stroking with his arms.

After the 15-meter start, the competitor has to move himself through the water by swimming. As we said before, the trick here is to generate as much power as possible using good technique that minimizes drag and converts as much of that power as possible into forward motion.

The last five meters is the finish. The swimmer makes his final stroke the most powerful he can and then glides forward with an arm outstretched so he can touch the timing pad with a fingertip.

The greatest all-around swimmer in the world right now—maybe ever—is Michael Phelps. But the *fastest* swimmer is France's Frédérick Bousquet. On July 22, 2009, he broke the world record in the 50 free and became the first man to crack the 21-second barrier.

Sometime around the 20-meter mark, after the initial speed generated by the diving start had dissipated and Bousquet was running on his own swimming power, he was moving at 5.3 mph. In terms that weekend athletes will be able to relate to, that equates to a little over eleven minutes per mile, which is faster than most people jog.

The 50-meter free is a sustained explosion of power, an all-out effort to cover the distance in as little time as possible. Some racers don't even bother to breathe during the event, in the belief that it wastes more energy than it generates. Some are even convinced that the expansion and contraction of the chest during breathing compromises hydrodynamic efficiency, the ease with which the body slices through water.

That might sound a little over the top, but we're talking about a race in which the winner can be decided by who cut his fingernails shorter that morning. Even though the official timing results are given in hundredths of a second, they're actually measured to the thousandth, which is why two competitors can have the same official time even though one of them finished first and the other second. How fast is a thousandth of a second in human terms? About four hundred times faster than the blink of an eye. (The only Olympic event with official results stated in thousandths of a second is the luge.)

With margins like that separating gold from silver, it's little wonder that swimmers look for every possible advantage, no matter how seemingly slight, right down to shaving off body hair. Mark Spitz, the legendary Olympian of the 1972 Munich Games, was thought to be the exception in that school of thought. He told Russian coaches that his moustache made him faster because it cut through water more efficiently than his bare lip would. Hard to argue with a man who broke seven world records on his way to seven gold medals, but he was just kidding. However, he did swim without goggles.

The detailed physics of swimming can get a little complex, but the basics are quite simple. In order to move through the water, you first have to generate force to get the mass of your body going. If you've ever been on ice skates, you know that a single modest push-off with one foot can send you gliding across the ice for the length of a football field. But push off the wall of a swimming pool with both legs as hard as you can and you'd be lucky to get twenty yards, even though you used four or five times more force. The difference, of course, is the amount of resistance your body encounters after you get moving.

So the competitive swimmer has two challenges in order to go as fast as possible: The first is to generate as much propulsive power as he can, and the second is to try to minimize resistance so that more of that power goes into moving him forward instead of just shoving water aside.

A swimmer's propulsive power comes primarily from the arms, which move the hands. As the hands shove water back, the body moves forward, in much the same way as oars move a rowboat forward by pushing water backward. The propulsive force generated by a world-class swimmer is quite great, but so is the drag of water on his body. If you take that body out of the water and put it in a sleek racing shell, the athlete could easily hit 14 mph, because very little of the boat touches water.

The swimmer also has to contend with the drag his legs cause. Very little propulsive power comes out of the legs compared to the arms; most of it just goes into compensating for the drag they create. This is the reason that great swimmers have short legs: The drag of the legs is greater than the propulsive power they can create, so the shorter the better. (Rudy Garcia-Tolson, a double amputee who races in the Paralympics, looked like a torpedo as he won a gold medal in Beijing. His 2:35 performance in the 200-meter individual medley would have beaten four men, and seven women, in the 2009 World Aquatics Championships.)

There are several ways in which a swimmer tries to overcome drag. One is to use a stroke technique that makes his body stay as high on the water as possible. The more of his body that's out of the water, the less the water can hold it back. Another way is to make sure that his hands knife into the water as he reaches forward for the next stroke instead of inadvertently pushing forward, which is like stepping on the brake. He also breathes (if he chooses to breathe at all in the 50-meter free) by rolling slightly instead of picking his head up clean out of the water and plunking it back down. All of this is going on while he's also trying to keep his body in as straight a line as possible, without flailing limbs messing up his smooth path through the water. Because swimming is a less efficient means of propulsion than running, and because it's harder to push through water than air, a swimmer has to use four times as much energy as a runner to cover the same distance. So anything he can do to decrease drag will result in more forward movement for the same energy expended. Technique is critical to maximizing efficiency, and good technique has been shown to be of more benefit to a competitive swimmer than his VO_2 max, a measure of oxygen uptake usually of great importance to racers.

There are some tricks, too. You'll notice that when freestyle swimmers dive in the water at the start, or when they push off the wall in longer, multilap distances, they don't come up right away and start stroking. Instead, they stay below the surface and undulate their bodies the way dolphins do. This allows them to exploit and enhance the speed of the entry dive or push-off without all the inefficient splashing around involved in the conventional swim stroke. This technique is so effective that the international governing body for swimming has set a 15-meter limit on the distance competitors can stay down and do the dolphin kick.

There's only so much training swimmers can do to make themselves stronger and improve their technique. That's why they look for ways to reduce drag that are "free," i.e., take no extra effort. Shaving off body hair is one of those ways that costs nothing in terms of training effort. So is wearing a swim cap and goggles and leaving off the jewelry. Anything extra that hangs in the water can impede progress.

Which brings us, inevitably, to one of the fiercest sports controversies of the budding twenty-first century: the high-tech, full-body swimsuit. A great deal has been written about these miracles of modern sports science, much of it wrong or confusing. Let's take a look at them and see what kind of impact they've had on the sport and, potentially, our efforts to find the perfection point of the 50-meter free.

In February 2008, the world record for the 100-meter swim stood at 47.84 seconds. In the previous thirty-six years, it had fallen an average of .09 seconds per year. By the time the Beijing Olympics ended six months later, it had been broken five times and had fallen an average of .13 seconds *per month*. What happened in February that made the difference? Did the athletes suddenly get stronger? Was there some amazing breakthrough in technique?

The athletes didn't get stronger, and there were no breakthroughs in technique, because records fell in every type of stroke across every distance, and there surely couldn't have been that many different breakthroughs in a single month, or even a single decade.

There's no mystery about it. What happened in February was the introduction of a piece of technical wizardry known as the Speedo LZR Racer, a full body-length swimsuit that did for swimmers what turbochargers did for hot rods: instantly make them go faster for the same amount of available energy. The sporting and mainstream media reported on the suits endlessly, marveling over how Speedo engineers had managed the feat. One magazine focused on the suit's ultra low weight. Another waxed poetic over the welded seams that eliminated the drag caused by conventionally sewn seams. A major newspaper made much of the fabric itself, which was supposedly modeled on the water-repellent skin of a porpoise.

All those facets of the new suit were true, but they weren't what made it so disruptive to the sport. The real secret of the LZR is that it consists of a series of carefully shaped panels that push, squeeze, and compress the entire length of the wearer's body into a much more streamlined shape than the one he or she started with, and also keeps momentary bulges of skin, fat, and muscle from jutting out into the water.

As taut and lean as an elite swimmer's body looks, it doesn't look that way to the water it's trying to move through. It's difficult to see in swimmers, but if you've ever seen slow motion video of runners, you'll know what I'm talking about. There are body parts bouncing and jiggling all over the place. It's most obvious in the face, where cheeks droop with every step, making the runner look like an astronaut in a centrifuge. Look elsewhere and you'll see bellies jiggling, breasts and butts bouncing, thighs flapping around like hot water bottles, and arm muscles dancing as though they weren't connected to bones at all.

The same things happen in the water, but the consequences are much more pronounced. Any time a muscle or loose section of skin bulges or shifts, it's going to block the smooth flow of water and impede the swimmer's forward motion. You'll understand the effect if you've ever paddled a canoe. If you stop paddling and glide quietly, you can make the canoe turn just by sticking your hand in the water. You increase resistance on one side of the boat, slowing down that side while the other side keeps going, causing the canoe to slew around.

That's what happens to a swimmer when various body parts naturally flap around as he strokes. The LZR suit holds all those bits tightly in place and stops them from sticking out into the water and increasing drag. At the same time, it changes the overall shape of the swimmer's body into a more streamlined configuration.

The suit comes with some costs. One is the dollar cost, at $550 a pop. Another is physical: The suit is so tight that it takes half an hour—literally—to put on properly. Once it's in place, all that squeezing makes breathing more difficult, and it's so uncomfortable that the first thing wearers do when they get out of the pool is start tearing it off.

At first glance, the LZR suit seems an affront to the spirit of the sport. Swimming, unlike NASCAR and soap box derbies, is supposed to be about pure athleticism, not who's got the better gear. And while $550 may not seem like much money to well-funded teams from Europe and the United States, what about Third World entrants who can barely scrape up the scratch to take off from work and get themselves to the competition site?

International swimming officials brushed aside the criticism. Their attitude seemed to be that the new suit was legal because it didn't add anything to the energy provided by the swimmer. All it did was reduce friction, allowing more of the athlete's energy to be devoted to forward motion. How was that any different from a cyclist hunching forward to reduce the sail-like effect of his body or a downhill racer waxing the bottoms of his skis or a runner using 110-gram shoes instead of 150? And as for the suit's price, if you can't afford it, don't buy one.

Nobody mentioned that the hundreds of millions of people who watch swimming only once every four years love to see world records fall. To do that, they need to stay glued to their television screens. "Follow the money" comes to mind as an underlying rationale for allowing the new suits in official competition.

What happened at the Beijing Olympics in 2008 set off shock waves not only in the world of swimming but throughout all of sports. It's easier to talk about which Olympic swimming records were *not* broken instead of which were. Only two were left: Ian Thorpe's 400-meter free and Inge de Bruijn's 100-meter butterfly, both set four years earlier in Sydney. Aside from those, Olympic records were cracked sixty-five times. But even that staggering figure doesn't tell the whole story: In many events, multiple swimmers in the same race surpassed the old marks, but only one was counted as a record. Michael Phelps alone set eleven new records, eight of them in individual events. In the 200-meter butterfly, he set new Olympic marks in the heat and semifinal, then topped it off with a world record in the final.

To say that these jaw-dropping numbers represented improvements in athletic ability wasn't a position anyone was willing to take seriously. Even worse, all of those records were set with the original Speedo LZR which, within a scant few months, would already become obsolete. More advanced suits, most notoriously the Jaked, would cause even more records to fall in post-Olympic meets.

Once the LZR was declared legal, what happened next was much like what happened when the Supreme Court decided that customers couldn't be forced to use only telephones supplied by the phone companies: The floodgates opened and a tsunami of wondrous new gadgets poured out. Suit manufacturers all over the world scrambled onto the bandwagon with all sorts of ideas that pushed the boundaries of the possible. One maker discarded fabric altogether in favor of polyurethane, which is buoyant and lifts the swimmer higher in the water. Another figured out a way to trap tiny air bubbles in the skin of the suit, achieving greater buoyancy, although they denied it vehemently.

Apparently, they overdid it. At the world championships a year after Beijing, the world swimming authority reversed its earlier decision and announced that full-length body suits would be banned altogether. It also mandated a return to fabric rather than polyurethane. The only controversy that remained was whether the records set since the introduction of the LZR would be allowed to stand or be noted in the books with an asterisk. An asterisk, loosely translated, means "of dubious merit." If you see one next to a season home run record in baseball, it generally means the player was later found to have been on steroids. See one in the swimming books and the athlete was wearing the swimming equivalent of a James Bond jet pack.

One good thing is likely to come out of this debacle: Elite swimmers now have their eyes set on breaking technology-assisted records wearing only ordinary swimsuits. It's going to take a while, but when it happens, fans of the sport can take comfort in knowing that the athletes did more to earn their places in history than plunk down money for an outside assist.

Needless to say, in our quest for the perfection point in the 50-meter free, we're going to assume that the athlete is wearing a conventional swimsuit and relying only on his own abilities to get him to the finish line.

There are other external factors that contribute to a fast swim, aside from high-tech supersuits. The most important is the pool. You hear swimmers talk about "fast water," meaning that a particular pool allows swimmers to post better times. It's not just in their heads: The effect is quite real.

It starts with the depth of the pool. Swimmers generate waves as they churn down the lane, not just at the water's surface but below it as well. These waves travel rapidly down to the bottom of the pool and then bounce, in the same way that a sound wave echoes off a wall. The returning wave creates turbulence that slows the racers down. The deeper the pool, the more these waves will be dampened on the way down and up, resulting in a smoother and therefore faster ride for the swimmers. Modern competition pools have a uniform depth of seven to nine feet. (There's no extra-deep diving end like the one at your local Y, and no shallow kiddie end, either.)

Waves travel sideways, too, affecting swimmers in adjacent lanes. One way to ameliorate this effect is to make the lanes wider. While standard swimming pools have seven-foot-wide lanes, high-end competitive pools can have eight-foot or even nine-foot lanes. Lane lines, the cables strung the length of the pool and studded with plastic floats, play a role as well. They aren't just there to keep the swimmers apart. They neutralize waves that try to travel from one lane to the next. In addition to new designs, the organizers of important meets sometimes use double lane lines to further dampen waves.

Even more important than the lane lines are the gutters at each end and along the sides of the pool. As we said, swimmers create waves as they move. In short, high-speed events like the 50-meter or 100-meter free, these waves can achieve considerable height and strength. When they reach the end of the lane, they rebound and begin traveling in the opposite direction. One of the reasons racers dolphin-kick their way below the surface after a turn is to make sure they get under the wave instead of getting hit by it at the surface. Gutters are designed to "swallow" as much of the wave as possible before it has a chance to bounce away. The better and more expensive gutters literally remove the water in the wave from the pool altogether, and return it through outlets at the bottom of the pool so as to minimize disturbance. Gutters along the sides do the same thing, to prevent lateral waves from hitting the swimmers. In some high-end pools, such as the one at Beijing, there is an extra lane on each side, which remains unoccupied during a race. Its only function is to give lateral waves a chance to dissipate as they bounce.

Even the design of the touch pads at the finish line figures in. The best ones hang from the gutters and don't rise above the surface. Seems obvious and easy, but manufacturers put their brand names and logos on the pads, and they want the top part to stick out of the water so they can be seen on television. This blocks the gutter and prevents it from doing its job, which is to quell wave action.

The temperature of the water makes a difference, too. Overly warm water saps a swimmer's energy, and cold water can shock the system, slow down muscles, and even cause shivering, which wastes huge amounts of energy. The ideal temperature is 78–80°F.

Then there are the athletes themselves. If you watched swimming during the Olympic telecasts, you might have noticed that the bodies of the athletes bear a striking resemblance to one another. You can train twice as hard as the next guy, but if you don't have the right build, you're eventually going to hit a wall (so to speak) and never make it into the elite ranks.

The best way to describe the ideal athlete in any sport is to look at the one who comes closest to the optimum. When it comes to swimming, the hands-down favorite is Michael Phelps.

The first thing you notice is the length of his arms. Leonardo da Vinci set out to demonstrate that an ideal man's "wingspan" is equal to his height, which he did in this famous drawing known as "Vitruvian Man." That proportion may be ideal for the ancient Roman architects who postulated it, but for a swimmer, it won't do at all. Phelps's wingspan exceeds his height by three inches, giving him extra reach and the ability to push more water backward during a stroke. His torso is also very long in proportion to his legs, which allows him to ride higher on the water. His short legs (Phelps's inseam is only 32 inches) in turn produce less drag, and his size-14 feet mounted on hyperflexible ankles act like flippers, giving him a very powerful kick.

Phelps didn't swim the 50-meter free when he won eight gold medals in Beijing. The training is so specialized that it would have compromised his other events. But there's reason to believe, were he to give the 50-meter a try, we'd have a whole new perspective on that event. Whether he does or not, Phelps is still the best model we have for getting to the ideal 50-meter sprinter.

At least on the surface. Below the surface, muscle composition is another factor. Swimming power and especially upper body strength are crucial to success in sprint swimming—86 percent of a racer's performance in a freestyle sprint event results from muscle strength and the ability to develop power. The rest comes from efficiency and overcoming drag. (For the competitive distance swimmer, the strength component is less. At 100, 200, and 400 meters, the contribution of muscular strength drops to 74, 72, and 58 percent, respectively.)

Studies have shown that, in swim races over 100 meters, the mix of fast-twitch versus slow-twitch muscle fibers doesn't make much difference. Any slight differences are buried under variations in training, stroke mechanics, and racing ability. The 50-meter free, however, is the one swimming event in which a preponderance of fast-twitch muscles is an advantage, so that's what we're looking for in our ideal swimmer.

One of the most hotly debated issues in the world of competitive swimming today is whether anyone wearing a conventional swimsuit is ever going to break Frédérick Bousquet's world record of 20.94. Consider that Alexander Popov, described by many experts as the most perfectly constructed sprint swimmer ever, managed only 21.64 without the speedsuit. Given that the world record dropped only six tenths of a second in the twenty years before the new suits were introduced, the seven-tenths margin between Popov's time and Bousquet's is a seemingly unbridgeable gulf.

Back in the late 1980s and early '90s, outstanding sprinters like Matt Biondi, Gary Hall, Tom Jager, and Anthony Ervin were all turning in times in the 22.0 neighborhood. According to 2009 U.S. World Championship Team Head Coach Sean Hutchison, it was a kind of golden age of sprinters.

"Those guys were amazing," Hutchison told me, "and they happened to cluster in the same era. I don't see today's sprinters having the same capability, and I don't believe we'll see a steady, linear improvement toward 20.0 without the high-tech suits. There's going to have to be some kind of major breakthrough in technique to get there."

Why should that be? What if we just assumed that our perfection-point sprinter was extremely tall? Olympic swimmers take an average of 36 two-arm strokes per 50 meters. This equates to about 8 feet per second, not taking into consideration the start. Most of the current and former record holders are between 72 and 78 inches tall. If the tallest man in the world could swim at that pace, he'd cover approximately three feet more per second and could finish three seconds ahead of the world record times, somewhere between 17 and 18 seconds.

Unfortunately, it's nowhere near that simple, because swimming is about much more than just generating power. Terry Laughlin, founder and head coach of Total Immersion Swimming, would probably agree with Hutchison's notion that a breakthrough in technique is required.

"What made Popov so great," Laughlin contends, "wasn't intense training or exceptional power. It was his stroke efficiency." Laughlin points out that Popov generated considerably less power than his competitors but outswam them anyway.

Laughlin is not alone in his thinking. Rick Sharp and Jane Cappaert of the International Center for Aquatic Research studied 100-meter-freestyle swimmers at the 1992 Olympics and discovered something remarkable: The men who made it to the finals averaged a 16 percent lower power output than the swimmers who didn't qualify.

Most sprinters train by going as hard as they can against the clock and trying to improve their times. The secret training regimen devised by Popov's coach Gennadi Touretski was very different: He'd have Popov swim as hard and as fast as he could, but only while maintaining perfect form. The instant Touretski detected any deviations from the standards of perfection he'd devised, he'd slow Popov down. The amount of training on any given day was determined by how long Popov could swim without his form breaking down, even if it meant that he'd be putting in fewer miles than convention dictated.

Here's how Russell Mark of USA Swimming put it: "As in most sports taking place over a period of time, efficiency is critical to achieving success in swimming because swimmers are always trying to find that balance between generating speed and the amount of energy expended to do so. Efficiency is governed by technique because the swimmer is always battling the forces of the surrounding water as they're trying to move through it. Alexander Popov is highly regarded to be the epitome of efficiency. He made swimming the 50-meter and 100-meter freestyle—events that are typically thought of as the raw sprints dominated by strong beasts that can rip and thrash through the water—look so easy and so smooth. He was the sprint king, but his technique and efficiency made it look more like he was swimming a 200-meter race instead of the shortest distances."

So determining the perfection point is not just a matter of power, and we're not looking for an especially muscular specimen to swim the ultimate 50-meter free. We're looking for someone who has an ideal swimming physique and great power but who can utilize that power while maintaining perfect technique.

To help determine the perfection point, I turned to world-renowned orthopedic surgeon and expert in sports physiology Dr. Benjamin Domb. His starting assumption based on several widely held principles is that we have an athlete who can produce 500 watts of power, of which 10 percent can be funneled into propulsion (500 watts is about two and a half times the power you use in climbing a set of stairs). The most straightforward way to derive the maximum speed for an ideally built swimmer is to use the same equations that determine "terminal velocity" for a skydiver, which is the maximum speed he can reach in free fall without opening his chute. Simply stated, it's the speed at which the air pushing back at him is exactly equal to the force of gravity pulling him down. When that speed is reached, he will stop accelerating and fall at a constant rate, which is about 220 mph. For a swimmer, terminal velocity is the speed at which the propulsive force he's generating is moving him through the water as fast as it possibly can because there's nothing else he can do to reduce his drag any further.

Drag is related to how much water his body has to push through. The best way to visualize this is to imagine a photographer in front of the swimmer aiming a camera directly at him and then creating a silhouette. The amount of area occupied by the silhouette is known as the swimmer's cross-section, which you can think of in aircraft-designer terms as "how much body he shows to the water." The less cross-section, the less drag, which is the "secret" science behind how the LZR Speedsuit works: By squeezing various parts of the swimmer's body, it effectively reduces his cross-section (which we designate as A), thereby reducing the amount of water he has to push through and, thus, minimizing his drag.

There are several other factors that go into the equation, such as the swimmer's mass, his propulsive (or accelerative) force, the density of water (rho) and drag coefficient (C). That last bit is pretty complicated in itself, consisting of several components such as form drag, wave drag, and frictional drag, but those can be consolidated into a single term.

The equation looks like this:

$$\text{Terminal Velocity} = \text{Sqrt}[(2 \times \text{mass} \times \text{acceleration}) / (C \times \text{rho} \times A)]$$

Plug in the values we came up with, and it looks like this:

Fastest speed = Sqrt[(2 × 91 kg × 20) / (.4 × 1000 × .5)

That works out to 4.26 meters per second. However, this isn't the final answer. It's just an absolute limit for the body moving through water at maximum power output. Realistically, a human athlete will never be able to utilize more than 20 percent of 100 watts of power. Plugging that back into our equation yields an absolute upper speed of 3.8 meters per second.

A 200-pound athlete with perfect form generating that kind of power could do the 50-meter free in an astounding 18.15 seconds. And nobody will ever better it.

Update: On December 18, 2009, César Cielo of São Paulo, Brazil, swam the 50-meter long course in 20.91 at the Brazilian Championships, breaking Frédérick Bousquet's world record by .03 seconds.

CHAPTER FOUR

PERFORMANCE-ENHANCING SUBSTANCES

A digression

Speculation about the limits of human athletic achievement inevitably leads to a discussion about performance-enhancing substances. While weight lifting is our obvious takeoff point for this digression, the topic is applicable to every sport we cover in this book. But if you're expecting the usual rant about the horrors of steroids and the evil athletes who use them, you might be in for a bit of a surprise.

In this day and age, it's pretty much impossible to talk about the perfection point of humans and not acknowledge the role that steroids have played in complicating this discussion. In science you never assume anything; all your variables are accounted for except the one you're trying to understand. Steroids are a difficult variable to take into consideration, because in some cases they really are an unknown—the only people who we know for sure have used them are the ones who got caught. Everyone else could be that good or just that good at not getting caught. The mathematical equations that we use to calculate perfection points don't have a variable for this uncertainty.

But it's not just a matter of who uses and who doesn't, who's good and who's bad. This is not a classic tale of good versus evil in which everyone is dressed in black or white. In this saga, everyone wears gray.

First, in case you're wondering whether performance-enhancing substances really work, the answer is an emphatic and unambiguous *Yes*. The improvements can range from the subtle to the truly immense. They can chop a few hundredths of a second off a sprinter's time, making the difference between an Olympic gold medal and obscurity. They can turn a journeyman baseball player hitting fifteen or twenty homers a season into a superstar who hits forty, or vault a Tour de France cyclist from the back of the pack into the spotlight of world fame. Performance-enhancing drugs really work, which is why athletes in the upper echelons routinely risk their careers and their medals to use them.

Are they risking their health as well? Without question some of these substances are harmful, although there's a pretty good argument that in small, "reasonable" doses, the side effects are negligible. This gets us into the question of what a performance-enhancing substance actually is, and this is where things start to get hazy.

Everything an athlete ingests is a performance-enhancing substance, including spinach, milk, and bread. All contribute to muscle growth, bone strength, good circulation, and every other aspect of physical health that all of us strive for. In addition to the kinds of things that help athletes and the rest of us gain general fitness, there are chemicals that have more immediate effects on performance.

As an example, take sugar. In its various common forms, such as glucose, fructose, and sucrose, sugar is a simple, fast-acting carbohydrate that can supply short-term energy to depleted muscles. Get some into your body at about mile 20 of the marathon and it just might make the difference between a personal best and a total meltdown.

Then there's caffeine. That jolt you feel after knocking back a few espressos might help you jump-start your day or keep you alert for an all-night cram session. If you're an athlete, it might allow you to motor your way through a 5K race ten or fifteen seconds faster than you normally could.

Notice that I said "normally." Is coffee abnormal? If you scarf down a cuppa before you step up to the starting line, are you cheating?

As a matter of fact, no, you're not cheating. Even though caffeine is a powerful nervous system stimulant that has proven performance-enhancing capabilities, it doesn't appear on the banned list of substances of any athletic governing organizations, including the International Olympic Committee. Neither does sugar, in any form, including the little gel packs that provide it in quick-digesting liquid form and are handed out at aid stations by many race organizers. Most of those packs also contain caffeine, which, unlike sugar, doesn't exist naturally in the body. By any definition, caffeine is a decidedly unnatural drug, and it's far from safe: In high doses it can cause palpitations, acute anxiety, hyperkinesia, and a whole host of other nasty side effects.

So how come caffeine isn't frowned upon the way so many other substances are? I have no idea, other than that it would be difficult to ban coffee.

One of the more interesting chemicals that appears on the banned lists is testosterone, the naturally occurring hormone that is responsible for a wide range of male characteristics, including body hair, a deep voice, and strong muscles. Enhancing your natural supply of testosterone can increase muscle development, so it's an obvious attractant for athletes. Sports governing bodies don't like that idea, but because everybody has testosterone, including females, it can't be banned as such. Instead, the governing bodies have established limits on how much of it can appear in an athlete's blood. The assumption is that, if these limits are exceeded, the athlete must have taken some in contravention of the rules, and he or she is presumed to have cheated.

Seems reasonable at first blush, but think about it for a few minutes and you'll begin to see that there are a boatload of problems with this approach. The first is determining how much is "normal." In the Olympics, for example, all athletes are tested, regardless of which sport they're in. (Three years ago, some chess organizations began administering drug tests to their members in anticipation of chess eventually becoming an Olympic event. The fact that none of the drugs on the banned list could possibly enhance chess-playing abilities didn't matter: A rule is a rule, and even equestrians—and their horses—are drug-tested.) Yet it's a fair bet that the kinds of competitors attracted to weight lifting and boxing are going to have much higher testosterone levels than those who favor dressage, a kind of ballet for horses. So the objective for athletes isn't necessarily to forgo artificial testosterone: It's to see how much they can inject without going over the prescribed limits, which kind of makes the whole affair a little ludicrous.

Even worse, the standards themselves are arbitrary and have little scientific credibility. Basically, the numbers are a best guess of the highest testosterone level someone could naturally have. No large-scale surveying or testing was done to establish what that level should be or to see if it might unfairly exclude some portion of the population. That someone cheated is therefore a *presumption* without hard proof.

That was the case made by Floyd Landis, the 2007 Tour de France winner who had his title questioned when his testosterone level exceeded allowable limits. He claimed that he never ingested the stuff and that his level was naturally high. All of his appeals were denied, and his title was permanently stripped. (Landis's case was only one in a very long string of cycling drug scandals. The Tour de France may be the world's leading laboratory for the development of new and hugely creative approaches to better racing through chemistry.)

Whether what happened to Landis was fair or not—even though he eventually confessed to years of banned substance use, as of this writing he still denies taking testosterone—it did bring to public attention some of the problems attendant to the worldwide antidoping effort. And there are others. As an example, athletes with certain medical conditions can get waivers allowing them to use substances on the banned lists. While seemingly compassionate, that policy is also unfair to athletes who compete against the people who got waivers.

One of the things that gives people a lot of comfort that the tests are fair is the fact that nearly all of the superstars who were accused of cheating eventually confessed, even if it sometimes took years. Some claimed that they were given steroids by their handlers and took them unknowingly, but a lot just came clean and owned up, usually after they had no choice because the evidence was overwhelming (what I like to call the "stained blue dress" confession). Most notorious among the latter was Marion Jones, the track-and-field princess who vehemently protested her innocence but eventually did jail time for her blatant transgressions.

By the way, caffeine actually does appear on banned lists, as an "in excess of" entry. Drink two cups before your 1500 meter run and you'll be OK. Suck down a whole pot of dark roast and you could lose your title and be banned from the sport altogether.

The use of steroids and other performance enhancers says more about the nature of the business of sports today than it does about the character of the athletes. We look at baseball players trooping past congressional hearings by the dozens and cluck our tongues at those reprehensible traitors to all that is good and pure about the great American pastime. How can these role models for our children behave that irresponsibly?

Let's put aside for a second the question of whether you really want to use your typical baseball player as a role model in the first place. After all, would you want your kid growing up to be like Babe Ruth or Barry Bonds or Ty Cobb? Instead, let's look at what drove those superstars to do what they did. I'll start with a confession of my own.

I don't drink or smoke or take illegal drugs and I'm 100 percent confident I never will. But I'm an athlete, and even though I'm an amateur, I'm pretty competitive, and I understand the competitive spirit.

I try to imagine what it must be like for some kid from Iowa who's dreamed about being a professional baseball player since the first time someone put a ball in his hands. His father was the same way, but any thoughts the old man had of a big-league career got derailed by an undersize frame and an utter inability to weigh more than 160 pounds.

Theirs being an absurdly functional family, the two spent countless hours working on the kid's skills, going to games, talking baseball, and pretty much wallowing around together in their shared love of the game. Dad never pushed his son and in fact had to physically haul him off the field two or three times a week just to get him to eat something.

The kid got fine grades in school, but he was also a power-hitting utility infielder who made All-County when he was just a sophomore. He loved school and wanted to go to college, but when he got offered a one-year contract by a big league team, he wanted that even more. Dad put up some token resistance, something about getting an education and the odds of ever getting called up to "the Bigs," but he pretty much popped every button off his shirt when the contract was signed.

So the kid goes off to a minor-league farm team, where he's absolutely, utterly, insanely determined to work his ass off and get his shot in the Show.

Just like every other kid in the farm system. Many of whom are at least as talented as he is. Many who are more so.

The kid does pretty well. Everybody does pretty well. Scouts are not stupid, and they don't pick recruits they don't think will do well. Back in East Nowhere High, the kid was a star; here in camp, he's just another journeyman with a couple of potential strengths. He seems to have a flair for shagging line drives, but more important, he's got a really good eye at the bat. The coaches talk about how he seems to read the pitch before it's even fully thrown, as though he knows exactly where the ball will cross the plate, in plenty of time to do something about it.

The "something" he does about it, though, isn't much. Not too surprising, since he's—did I mention?—a string bean like his old man, weighing in at barely 170 pounds.

The coaches sure do like his work ethic, and his attitude is about as good as they've seen, but gee, we don't know . . . the fans could give less of a hoot about his work ethic. They don't care if you're a crack-addicted bank robber as long as you can reach those fences.

The kid is no dummy. He senses all is not well. His buddies keep getting called up while he hits the weight room and knocks himself out trying to put on some muscle. Like that guy from Kansas, you know, the 180-pounder who puttered around the farm system for two years hitting .262 and eight homers and then showed up this year at 210 pounds and started blasting balls out of the park like they were made of Flubber and he got called up to the majors and everybody pretended not to notice that he had to shave three times a day.

Anybody for whom steroid use in baseball is a big surprise might also be interested to know that the Lindbergh baby is still missing. Check this out:

Frank Robinson averaged 30 homers a year from age thirty to thirty-four. From age thirty-five to thirty-nine, it dropped to 22.

Babe Ruth averaged 46 home runs per year when he was twenty-nine to thirty-three, and it went down to 43 for the five years after that.

After Willie Mays turned thirty-eight, his average number of runs for the next five years dropped by a whopping 21.

"Nothing unusual there," you might say. "You get older, that kind of thing happens."

It used to. Barry Bonds was averaging 37 homers in the three years before he turned thirty-five. For the five years after that, his average jumped by an astonishing 15 homers per year. (In all of National League history only seven players have won the MVP Award at age thirty-five or older. Four of them were Barry Bonds.)

Mark McGwire's bump was even more amazing. In the first seven years of his career, he averaged 31 homers per year. In the last seven he averaged *49*.

Here are the numbers of home runs hit by Brady Anderson of the Baltimore Orioles each year from 1992 through 2001:

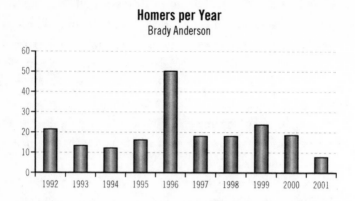

Homers per Year
Brady Anderson

Anything stand out? In case you need a hint, 1996 was the year Mark McGwire hit 52 homers, Barry Bonds 42, and Sammy Sosa 40. Speaking of Sosa, he hit 14 homers per year in the first five years of his career. Prior to 1998, he'd never hit more than 40 in one season.

From 1998 to 2001 he *averaged* 61.

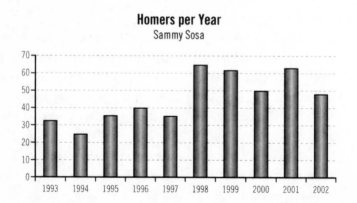

Homers per Year
Sammy Sosa

It's true that there have been many improvements in nutrition, training, and physical therapy that make it possible for athletes to squeeze out a few more years. But these numbers are ridiculous.

The kid from Iowa knows all this. He also knows that, for every superstar hauled before a hearing, there are dozens more sitting back in the locker room unidentified. So he goes home during a three-day break, and he sits down with his old man, and he says, "Pop, I'm working my butt off in Moose Breath Triple A while guys with half my talent are playing in Yankee Stadium and Fenway and making more money than this entire county. If I don't take the juice, I'm out of the game at the end of the season, so . . . what do you think?"

At first Pop is skeptical. His heroes are McGwire, Bonds, and Sosa. Surely they didn't . . .

But the kid is well prepared. "Yo, Pop," he says, and reviews our little history lesson about home runs per season. Just in case his father still didn't buy it, the kid had before-and-after photos of those three guys in his pocket.

So much for the illusion that baseball is clean. But still . . . "Gee, son, I don't know. Don't seem right somehow."

"It isn't, and I don't want to—hell, I haven't even had a French fry in fifteen years—but the thing is, unless you're some kind of freak, there's not much choice if you want to be in the game."

"What happens if you get caught?"

"Caught! Are you kidding me, Pop? Who's going to catch you? Nobody's even looking!"

"What about the players' union? Aren't they supposed to be protecting the players? You know, their health and whatnot?"

"Jeez, Pop, you gotta get out more. The players' union is who's refusing to allow drug testing."

But the man just won't give up. "Well, isn't it illegal?"

"So's spitting on the sidewalk. When's the last time you ever heard of a baseball player getting busted for unauthorized use of a perfectly legal drug? Besides, even when they get nailed for cocaine or drunk driving, they don't get thrown out of the game."

"OK, then." The old man is now ready with his trump card. It's his last one, but it's a good one. "What about your health?"

"Good question, Pop, so let me ask *you* one: If there'da been a pill thirty years ago could've put you on second base in Wrigley but maybe would give you some kidney problems or mess up your liver a little and there's no way on God's green earth you would ever get caught, what would you have done?"

How a genuinely honest man who truly loves his son would answer that question is the reason why steroids in sports suck. But so do a lot of other things. The harm done by steroids pales in comparison to other types of damage to players. The average length of a professional football player's career is less than three years. On any given day, hundreds of players are on the disabled list, and it's doubtful that athletes in any other sport, including boxing, live out their postcareer lives with the kinds of chronic physical infirmities that plague retired footballers. The average life expectancy for a man who played at least five years in the NFL is fifty-five. For a lineman, it's fifty-two. Steroids are hardly the worst health hazard in the sports business.

So here's my confession: If somebody had come up to me in my prime and said, "I've got a little something in a vial that the governing bodies haven't cottoned on to yet that'll give you enough of a boost to get you into the Olympics, but it'll take a few years off your life," I honestly don't know what my decision would have been.

And that's why I try to have a little compassion for professional athletes who take steroids. I sure don't admire it, but I understand it, and what I understand most of all is that none of them *wants* to take the stuff: They often have no choice if they're going to be in the game at all, because too many of their competitors are doing it and they're dead in the water if they don't keep up. The money is too overwhelming, the pressure too great, and the fact is that many of those who should be the ones putting an end to it are the ones who are looking the other way, including owners and managers, along with apparel companies who are marketing to kids.

Baseball is the perfect example. Back in 1968, the great pitcher Bob Gibson had a season ERA of 1.12. That means that, on average, he allowed just over one run per game. Now, while I love a pitcher's duel more than any other situation in baseball, the majority of fans want to see "offensive production," which means hitters hitting and getting on base. Someone like Bob Gibson was a threat to that concept, so what major-league baseball decided to do after his amazing performance was change the rules of the game. They lowered the height of the mound from fifteen inches to ten. Seems like a small thing, but baseball is a game of inches, and yanking five of them from under the pitcher made it significantly easier to hit his pitches. Hits went up, ERAs went up, and butts—paying butts—began coming back to the seats and the television sets.

Same thing happened when guys like McGwire and Sosa started banging balls out of the park and threatening Roger Maris's single-season record of 61 homers. The game soared in popularity, and money started flowing back to a league that had begun to look moribund. Home-run production grew so rapidly that there was talk about the ball being juiced, but it wasn't the ball, it was the players, and anybody with half a brain knew it. McGwire weighed 215 pounds as a rookie. By the time he started routinely knocking balls into the stratosphere, he weighed over 250 and had arms like Popeye. Did anybody really believe he'd just put in more hours at the gym in the off-season?

When a jar of androstenedione was spotted in his locker during a postgame interview, no punishment was meted out because there were no rules prohibiting its use. Androstenedione, which occurs naturally in the body and boosts the production of testosterone, could be purchased over the counter. And who wanted to upset the money train that was steaming along the tracks largely as a result of Big Mac's pursuit of the most revered record in the sport?

You have to wonder why McGwire and a host of other players felt compelled to hide their use of performance-enhancing drugs when it wasn't against the rules of baseball at the time they used them, and the sale of some of them was perfectly legal, too. (Andro wasn't banned by the federal government until 2005.) Were they embarrassed because they might be perceived as having gained an unfair advantage and therefore were less worthy of the accolades that had been heaped upon them? What about their teammates who used ibuprofen to ease aching muscles after—or during—a game? Nothing wrong with that. Did any of them use creatine, an over-the-counter substance even high school weightlifters take? What about any of a thousand perfectly legal, totally uncontrolled and unregulated supplements that promise a wide range of benefits, which, while largely dubious, still constitute an attempt to improve performance through the use of "artificial" substances?

I'm involved with a lot of athletic events and watch a ton of sports. My automatic assumption when I see people at the top of their games is that they did *something* that pushed the envelope to get there and stay there. It might be as clear-cut as shooting steroids or as questionable as overloading on some obscure root from a faraway rain forest that the authorities haven't discovered yet. Even disabled athletes at the Paralympics indulge in remarkable and shocking methods of boosting performance: Male wheelchair athletes have been known to boost their blood pressure and thereby their performance by using rubber bands to restrict output from their bladders. (I'll leave it to you to figure out how they did that.)

I don't know what the answer is. We certainly don't want obsessed athletes literally killing themselves to snag a few seconds in the spotlight, and they would: One reason the Olympics will never have an underwater swimming competition is because there are athletes who would rather risk drowning than come up short on the international stage.

The question is relevant in this book because we had to make some decisions as to how to handle the issue of performance-enhancing substances for the purposes of determining perfection points. Do we insist that our ideal athletes perform 100 percent clean? Do we assume that anything a juiced-up competitor can do could also be done by a nonuser if he worked hard enough?

Let's deal with the last question first, because it affects the first and is also a lot easier to answer. That answer is a simple no, and the logic of it is straightforward: Whatever an athlete in a strength-related sport can do clean, he could do better on the juice. No matter how big and strong his muscles get, steroids would make them bigger and stronger. The only exceptions would be those cases in which any further increases in muscle strength would threaten to break bones or tear tendons away. But bone strength and tendon resiliency vary greatly from individual to individual, so in hypothesizing our ideal athlete we can also make assumptions about his physiology that would preclude those limitations.

And now we're left with the first question: In determining the perfection point, do we insist that our athlete be free of steroids and hundreds of other performance-enhancing substances?

While I as much as anyone would like to see drug-free competitions so that the athletes we love will stop killing themselves to earn our admiration, for the purpose of determining the absolute limits of human performance, it's pretty much anything goes. This is by no means an endorsement of those substances, but to preclude their consideration would be to place artificial constraints on achieving perfection.

Suppose we were to conclude that the most weight a human will ever bench press is 1,000 pounds. When the debate on that begins—and the whole purpose of this book is to spark just such a debate—someone is very likely to argue, correctly, that we didn't take into account the possibilities presented by a plethora of potentially beneficial substances ("beneficial," of course, being entirely a matter of opinion). Durabol, a popular anabolic steroid, is currently on the list of banned substances for Olympic competition. But suppose someone were to discover that it occurs naturally in Brussels sprouts, and that eating a pound a day of that vegetable would have the same effect on muscle growth as a bodybuilder's regimen of 600 milligrams a week? Would we tell athletes they're not allowed to eat that much of the stuff? And how would we know if he did, or if it affects some people differently than others? What if new substances go on the banned list?

More interesting, what if others come off? Suppose we were to discover in twenty years that, much to the chagrin of so-called purists, there were anabolic steroids that had no significant side effects or that some of today's compounds were safe in lower doses? What if it turned out that it was actually safer to take small doses of those substances than it was to eat too much spinach or complex carbohydrates or protein-rich foods?

What if allowable limits of naturally occurring chemicals like testosterone had to be raised because we found out that too many completely clean athletes exceeded them? Or that there were ordinary foods that raised those levels? Or that the current, arbitrary limits were set incorrectly to begin with?

The fact is, we have no idea what's going to happen in the future and, furthermore, there's no basis for decrying the use of Durabol as "cheating" but allowing the free use of a stimulant like caffeine. As for imposing bans for the sake of the athletes' health, there are some pretty good arguments against that kind of manipulative paternalism, at least for adult athletes.

The first is that it's none of anybody's business how someone chooses to train, even if that training is patently self-destructive. If you want to see some voluntary self-destruction, spend an hour at a rugby training camp or, better yet, a ballet school. The number of shin splints, sprains, and dislocations among premier dancers would floor you, and it's not at all unusual for ballerinas to perform with multiple broken bones in their feet.

Another argument against the regulation of substance use by athletes for their own good is that it's hypocritical. We don't seem to have any problem subjecting high school football players to repeated injury, including the head trauma known as concussion, so why get all uptight if they pop a few pills to help them excel? How come we don't seem to mind boxers pummeling the daylights out of one another for the sole purpose of rendering the opponent too damaged to continue but we're shocked—shocked!—at the far less severe damage caused by steroids?

As I said, I don't like the idea of athletes using steroids, primarily because of the harm they cause when abused. But in trying to ascertain the limits of human performance, we can't rule out their use, because doing so places a barrier on achievement that is not only artificial and not only a moving target, but possibly a deceptive one as well. After all, can we really assume that the current record holders are "clean," not only according to present-day standards but to future ones as well? Does anyone really believe that all the steroid users have been caught and that all of the remaining champions are completely innocent?

Let's be clear: When we use today's top athletes as the starting point in our predictions of perfection points, we are *not* assuming that they're clean. We're not assuming anything one way or the other on an individual level. Collectively, however, some of them probably are using.

The question of performance-enhancing equipment is closely

related to that of performance-enhancing drugs. We've already shown how gear like the speedsuit and lifting shirt rendered records obsolete in swimming and weight lifting. The lifting shirt is an obviously ridiculous piece of equipment that ought to be banned outright, but the speedsuit is less obvious and has also been banned. Swimmers routinely shave off body hair; while allowable, is it legitimate?

Tough questions, but there's an easy answer, at least for our purposes: So long as the playing field is level, meaning that everyone is playing by the same rules with respect to equipment, we accept the rules of the sport as given, noting adjustments for performance differences unrelated to human capability. (As an example, our calculations for the longest golf drive and longest baseball home run assume still air and are easily adjusted for tailwinds.)

Also, to the extent that equipment rules are significantly altered, we may not even be talking about the same sport. Nobody compares driving distances in golf today to those of fifty years ago because titanium club heads, graphite shafts, and high-tech balls have changed golf into a different sport. The same is true of pole vaulting, cycling, and skiing, among many others.

Unfortunately, the distinctions aren't always clear. This is an area that people in sports are very reluctant to talk about. There doesn't seem to be much controversy anymore over whether the swimming records set in Beijing are comparable to the ones that came previously. Of course they're not, because the Beijing swimmers were wearing performance-enhancing suits that were unavailable to their predecessors.

Then again, the Beijing pool was an especially fast one, for reasons we discuss in the chapter on the 50-meter freestyle swim. Does that mean that, even had the speedsuits not been used, world records set in that pool should have been asterisked? The answer there would be a universal No, because the sport, as do all sports, recognizes that there will always be variations from venue to venue and from day to day. Sometimes it's possible to set limits on those variations, for example by setting maximum allowable wind speeds on the track to make records official. Sometimes the variations are judged too small to take into account at all, which happens to be the case with fast versus slow pools.

Baseball and golf are two examples in which wild variations are pretty much ignored altogether for the purposes of citing record performances. The "lowest score ever shot in a U.S. Open" is the type of dubious record that makes no sense at all when comparing one golfer's skill to that of his predecessors, because Opens are played on a different course every year. Baseball, the most statistics-happy sport on earth, isn't even played on a uniform field. Distances to the fences vary wildly from park to park. When the original Yankee Stadium opened in 1923, the distance down the left-field line was 280 feet, the center-field fence was 487 feet away, and down the line in right field was 294. By the time it closed in 2008, those distances were 318, 408 and 314. The old Yankee Stadium, along with six other major-league ballparks, wasn't even in compliance with MLB standards calling for a minimum distance of 325 feet to any fence. Yet we continue to completely ignore significant differences in field size when comparing home-run records. You might argue that it all evens out as players rotate from park to park, but what about the guys who play half their games in a home field that is either much shorter or much longer than most of the others? And how about Yankees from decades ago whose left-field fence down the line was a full 45 feet shorter than the MLB standard?

We've chosen in this book to play by the same rules as the governing bodies, ignoring distinctions where they're ignored by the sports themselves, taking them into account when the sport itself does.

But don't think it's going to end there, with just equipment and playing-field differences. The "next big thing" in athletic performance enhancement is going to be altering the body itself, using "genetic doping" or even surgery. Some swimming greats were rumored to have had toes that were naturally webbed, giving them an advantage in the water, so if a budding young Olympian decides to have his toes sewn together to turn his feet into flippers, is there any basis for an objection? Suppose he also elects to have two inches of bone removed from his femurs to reduce his leg length along with extensive plastic surgery to mold his torso into a more aerodynamic shape?

If you think things like that will never happen, I haven't done a good enough job getting you inside the minds of elite athletes.

CHAPTER FIVE

DRIVING FOR DOUGH

What's the farthest someone can hit a golf ball?

- -

Think golf is a contest between player and course? Think again. Golf is a contest between equipment manufacturers and the U.S. Golf Association. The manufacturers want to make clubs and balls that will let you hit a drive from a tee box in Buffalo to a green in Detroit. The USGA imposes rules to limit distances so the game is fair for everyone and golf courses don't have to be the size of Rhode Island. Within those rules, what's the farthest a human will ever hit a golf ball on level grass?

- -

Augusta National
Sunday, April 23, 2152

It wasn't going well. It certainly looked like it was going well, but it wasn't.

Julian Herrera had just birdied the fourteenth, the only hole at Augusta without a single bunker but with a green that looked as if a dysfunctional family of giant moles lived beneath the surface. Every putt was a crapshoot, a tap followed by held breath as the uneasy relationship between grass and gravity guided the ball the way air guided a knuckleball: randomly, and with precious little concern for hitting the target.

Julian had purposely kept his approach shot short and to the right of the pin, choosing to leave himself a longer and relatively flat putt rather than a closer-in but far trickier drop down a steep hill. He'd struck the ball with all the purity of a spring rain, and watched in satisfaction as the ball arced perfectly toward the nasty green, touched down and spun back the precise distance he'd intended, leaving him with an easy putt for birdie and the crowd with yet another memory etched by this blazing wunderkind.

At least that was the plan. What had actually happened was that he'd hit the simple wedge so purely that he'd overflown the flagstick, leaving the worst hill on the green between his ball and the hole. So far on this day, of the eleven other players who'd ended up in that awful position, three had two-putted, and they were the lucky ones: The rest needed three to get down.

Julian's caddie had watched all eleven putts on the paper-thin television screen taped to his man's bag. "You gotta whack it up top of the hill two full feet right of where you think you should," he'd advised, persisting in the face of his man's adamant head shake.

"I miss," Julian had whispered between clenched teeth, "that sucker's going off the green."

"You don't try," the caddie had shot back, "crowd'll turn on you like you'd starved your children."

"I don't have any children," Julian had responded dispiritedly, but he'd gotten the point.

Still an amateur, he was tied for the lead in the most prestigious golf tournament on the planet. No amateur had ever won it, and only seven had managed second place in the tournament's 218-year history. Julian knew that sportscasters the world over were at that very moment feverishly clicking their online thesauri in a pitched battle to see which of them could outgush the rest in their torrential praise of the phenomenal twenty-one-year-old, some of them daring to speculate aloud as to whether he could actually win. By any measure, even finishing in the top twenty would be a stupendous achievement, but Julian knew that, were he to make liars of his public admirers, they'd find a reason to fault him rather than admit they'd gone seriously overboard.

When he'd heeded his caddie's advice and sent his ball two feet farther right than anyone had so far dared, he'd groaned along with the crowd and the commentators as the traitorous little pill picked up speed on the downside and had no chance of turning into the hole.

Until it did. Picking his ball out of the cup and casually touching two fingers to his forehead in acknowledgment of the crowd's frenzied screams, Julian could just hear the on-air announcers ramping it up: "Genius . . . brilliant . . . game-changer . . ."

They'd talk about how Julian Herrera was a steely-eyed ball striker who could put himself into a zone of trancelike intensity and confident relaxation, that the kid's mental toughness was so extraordinary you could set off a firecracker during his backswing and it wouldn't affect him in the slightest.

That last part was true. The rest was horsepucky. Julian Herrera was an ouch-cube of jangling nerves and knee-rattling fear who could hit a hundred dead-perfect shots in a row and be fully confident that the next hundred would be just as good, and yet the terror of playing public never went away. Which is why, at least from Julian's perspective, things weren't going anywhere near as well as they seemed to outside observers.

It had always been thus with Julian, and whenever it got to be too much, he'd simply work himself to exhaustion. At first it was running until he dropped, then lifting weights, then riding bicycles or swimming. Local coaches had spotted him early on, and in everything he tried, he'd excelled. It was all so easy.

Then someone put a golf club in his hands. Julian, who'd seen plenty of golfers swing, wound up and took a whack at the ball with such a rush of naked power that onlookers caught themselves unconsciously taking a step backward lest they be hit with the resulting shock wave.

They needn't have worried. Julian had missed the ball completely and nearly fallen over from the effort. He tried again . . . same result. The third time, he finally managed to make contact, but it was so off-center and glancing that the ball skittered off the tee as though embarrassed.

A skinny kid from his third-period math class stepped up and took the club from him. "Not as easy as it looks, is it?" he asked, then took a seemingly effortless swing and sent the ball soaring.

Handing the club back, he said, "Trick is to set up, then swing so that you come right back to exactly where you started. Can't be moving your body around all over the place."

Julian had no ego when it came to improving himself. If someone had a better way and could prove it, he paid attention.

"You gotta stay centered," the skinny kid said.

Julian tried it, a few very light swings, the other kid watching in satisfaction as the best athlete in the state gradually began to understand.

"Once you groove that in," he said, "you start adding power. Every time you go out of control, back off until you get it back. Then swing harder again."

It was sound advice, and it made sense to Julian. Much of his athletic prowess had been based on the application of pure strength, but when he got into a swimming pool and tried to power his way across, he was easily overtaken by kids with half his strength. It was in the water that he learned the virtues of technique, and once he added efficiency to his power, he became unbeatable. As he swung the golf club during his first day on the range, it didn't take him long to see that even incredibly minute deviations from the perfect swing resulted in drastic reductions in distance and accuracy.

Four years later, Julian showed up at the world long-drive contest in Mesquite, Nevada, all six-foot-six of him, shrugging off the snickers from the veterans and walking off with the first-place trophy and the new world record.

At a small celebration in a local bar afterward, one of the veterans came up to congratulate him. "Impressive," he said. "But so what?"

At Julian's puzzled stare he added, "The rest of us, we're ex-football players, basketball players, couple'a golfers . . . this is a lark for us. You, you're a strong kid, sure, but now what? Think there's a living in hitting long?"

"What do you think I should do?"

"Try playing the whole game," came the answer. "With your length and control, you'll get closer to the flag in fewer shots and maybe get away with not having the world's best short game."

The "whole game" drove him nuts. He was used to pouring raw strength into his sports, and the need for finesse had never been much of a factor. He was always intimidated by audiences, but when he was pounding along a track or plowing through a pool, fear didn't really matter. Standing over a putt or a tricky wedge shot, it mattered a great deal. Golf was a game in which being off by a few millimeters could mean the difference between sticking it near the flagstick or dumping it into a lake. Even veteran professionals occasionally got the yips over a putt. Julian got them over *every* putt, but he'd learned how to confine them to his brain and disassociate them from what his body was doing. He didn't realize that, in doing so, he'd found the Holy Grail sought by every athlete who'd ever lived.

He was entering amateur tournaments all over the country, relying, as the veteran had suggested, on his prodigious length to get him close while he continued to hone his short-game skills. He'd also become a student of the sport, building the relevant muscles the way a bodybuilder sculpts his physique, one piece at a time using finely focused workouts. He'd identified which muscles did precisely what, what the optimal balance among them was, how to measure his progress and make midcourse corrections. He'd also made a science out of the equipment, figuring with precision the exact set of physical characteristics required to achieve maximum distance while staying within the letter, if not always the spirit, of USGA regulations.

He had it down, or thought he had, but what he'd never counted on was how many maddening decisions he'd be called upon to make during tournament play. Sure, on a good day he could whack a ball 400 yards with acceptable accuracy, and had gone even longer on the range, but knowing when to take the chance on powering a drive like that wasn't always obvious. If he was ahead, was it worth the risk to try to put a few strokes in the bag in case someone made a run from behind? If he was behind, should he protect his position and hope someone in front of him would fall away, or risk finishing up in obscurity by tossing the dice?

He didn't get any help from the gallery. They saw great golfers shoot great scores every day. What they wanted from Julian Herrera was a testosterone-fueled blastathon that would leave them gasping in awe, and he obliged them just often enough to keep them hanging and guessing. But he was still the one left to make the decision. If he guessed right, he was the brilliant hero wined and dined and lauded endlessly; if he guessed wrong, he was the goat, vilified by the same capricious pundits who'd hailed him as "the greatest ever" just the day before.

And now he'd birdied the fourteenth at Augusta to put himself into a tie for first with four holes remaining.

His caddie looked up from the screen taped to his man's bag. "Conroy just birdied twelve."

Julian, knowing dozens of cameras were on his face, tried not to betray his dismay. Jackie Conroy was one of the winningest golfers on tour and had won the Masters twice in the last five years. "What do I have to do to beat that guy?"

The caddie thought about it. "You gotta beat him. Not just play good golf. You gotta beat him."

Right. Exactly. "What the hell are you talking about?"

So the caddie told him about Franz Klammer, the 1976 Olympic downhill skiing gold medalist whose epic winning run was one of the most harrowing ever seen, the Austrian looking completely out of control and in imminent danger of crashing from the time he left the gate to the moment he crossed the finish line. Asked by a reporter why he'd risked his neck on a maniacal descent like that, Klammer had replied . . .

"To win, you have to risk losing." The caddie searched Julian's eyes as he said it, waiting for some sign of recognition. They were standing on the tee box of the number fifteen par five, "Fire Thorn," a 530-yard respite from the maddening tricks played on golfers on the rest of the holes designed by Alister MacKenzie. The fiendishly clever Scottish physician had made a careful study of camouflage techniques in order to protect field hospitals in the First World War, and had carried his knowledge over to the golf courses he designed the world over. Many of his tricks of perspective had been made obsolete by electronic distance finders but, as if in anticipation of that technology, McKenzie had planted plenty of basic challenges and exasperating decision points between golfer and hole.

The fifteenth was a good example. In terms of distance, the green was attainable in two good shots by many of the top professionals, most of whom could lay down a fairly accurate 290-yard drive and follow-up with a solid five-wood. The problem was the pond McKenzie had dropped right in front of the green, turning the second shot from a "Let's see what happens" percentage grab at an eagle into a potentially disastrous step off a cliff. Most of the tour players laid up in front of the water, banking on planting the approach close to the flag for a birdie chance. McKenzie, anticipating this, added a hillock in front of and to the right of the green, tempting the longest hitters to go for it in two because the little mound would help guide errant shots back to the green.

Julian knew what was in his caddie's mind. "You're saying I should go for it in two."

But, as it turned out, he didn't know what his caddie was thinking. The bag carrier shook his head, locking his eyes on Julian's.

Julian turned his head up slightly to gauge the wind, but got nothing in response. It was dead calm. It was also unnaturally dry for Georgia in April, which made the unseasonable heat more tolerable.

"You gonna hit?"

Julian had been barely aware of the player he'd been paired with for the day. He looked out over the fairway and saw the group ahead of them trudging toward the pond. "You go ahead," he offered.

The other player shrugged, teed up, and hit a perfectly acceptable drive down the center of the fairway.

Julian nodded an acknowledgment of the shot, not trusting himself to speak without croaking, his throat suddenly having gone dry. He made no move toward his own clubs, just continued staring at the fairway.

After a few uncomfortable seconds, the tournament official assigned to their group stepped toward Julian's caddie. "Is he planning to hit pretty soon?"

The caddie was also looking down the fairway. "Pretty soon."

The official turned toward Julian. "Sir . . . ?"

By way of response, Julian pointed. The official followed his gesture but saw only an empty fairway.

Puzzled, he turned back. "I must ask . . . what are you waiting for?"

Julian pointed again, and again the official turned. This time he looked toward the green over a quarter-mile away and saw people on it.

By this time, the players behind them had finished the previous hole and were walking up to the teeing area. Seeing no one on the fairway, one of them asked, "What's going on?"

The official bent to explain in hushed tones. Before he could finish, one of the golfers stood straight up and said, "Baloney."

Ignoring him, Julian's caddie pulled a driver from the bag and handed it over. "They're gone," he said, nodding toward the distant green.

Julian took the 48-inch club, a length ideally suited to his shoulder-to-fingertip distance of 28 inches. He put a hand at either end, then lifted it into the air and lowered it behind his back, bending side to side from the waist at the same time. He brought the club back over his head, let go of one end and stooped to tee up his ball, a model made especially for long hitters. It was of marginal utility for other players, since its exceptional abilities only came into play at higher club speeds.

By this time a larger crowd had gathered, and Julian tried to put them out of his mind. It wasn't hard, because he had plenty to think about, but he tried to push all of that aside as well.

The fifteenth had a slight bend to the right that generally called for a little fade, but Julian didn't care about that right now. He could see about half the green but couldn't see the flag because of the trees along the right side of the fairway, but that didn't much matter, either. This wasn't going to be exactly what you'd call a finesse shot.

He set up two steps behind his ball and took three practice swings in rapid succession, loosening up his muscles and reminding them of what was about to be required of them. Then he stepped up to the ball and took his usual stance. With one last look toward the green to put the exact direction in his mind's eye, he returned his gaze to the ball, took a deep breath, let it out slowly, and swung.

By the time Julian's driver connected with his ball, the club head was moving at 176 mph. The face at impact was precisely perpendicular to the plane of the swing, and it was the exact center of the face that hit the ball. That generated a maximum "smash factor" of 1.5, meaning the ball flew off at an initial velocity of 264 mph. The angle of the club face with respect to both the shaft and the ground imparted backward spin of 3,050 rpm and sent the ball into the air at an ideal launch angle of 16 degrees.

The crowd was too stunned to let out the usual whoops and hollers; a collective intake of breaths was all that could be heard as the ball soared into the sky. It disappeared quickly, becoming too distant to see, and it was impossible to predict where it would land.

Some seconds later, a handful of spectators noticed a commotion on the green. Hands shot into the air and the mass of people gathered there seemed to shift suddenly and begin shimmering in the heat. It took the accompanying sound a full second to reach the tee box, where Julian was still poised with the club over his left shoulder, not having moved since he'd hit the ball. At that distance, the spectators on the green sounded like a faraway jet engine. At first they didn't even realize where the ball had come from. Once they'd begun to figure it out the noise grew deafening, and it continued to grow, swelling impossibly as the realization began to settle in that they'd just witnessed history.

Julian Herrera's ball had come to rest at the back of the green.

Golf is one of the most popular competitive participation sports in the country. Some 26 million of us play at least a round a year and spend over $60 billion doing it.

Those are some interesting statistics considering that golf is fiendishly difficult, impossible to master and maddeningly complex. The rules are so intricate that they're the single most popular source of conversation on the golf course. (A half-eaten apple in a sand trap counts as a natural obstruction even if the closest apple tree is eight hundred miles away. Whether you can move a ball away from a hole made by an animal depends on whether the hole is something the animal lives in or dug just for fun. You can move a ball away from ice if it came out of a soda machine but not if it's frozen rain water. And those are easy ones.) Technique in golf is more important than it is even in swimming, which is why those seventy-five-year old men down at the local muni with DISABLED flags on their carts routinely kick the daylights out of brash and unsuspecting studs half their age.

Golf is like several sports rolled into one. The skills you bring into play to hit a drive are barely related to those you need to make a putt or a tricky sand shot. It's also a sport in which "trying harder" is the worst thing you can do to improve your game. If you're six places back on the last stretch of a 10K race, you can dig down and find that last ounce of strength and adrenaline that will propel you to the finish line. In golf, you have to do the opposite: force yourself to relax, which is an oxymoron right there, and forget about everything except the shot you're about to hit. Golf is the closest that most Westerners ever get to a Zen-like experience, which might explain why we're so bad at it. The average handicap in the United States hasn't changed in decades despite all those billions we spend on "proven, sure-fire" gadgets for improving it.

Like baseball, golf is a numbers-happy sport. There are statistics for every aspect of the game, some of them more sensible than others. Among the dubious ones are things like putts per round. Lousy golfers often have lower putts per round because they're going to be much closer to the flag when they finally get onto the green, whereas great golfers, who can hit the green from 200 yards out, generally end up farther away.

But there's one number that speaks for itself and doesn't need a whole lot of explication: the length of a drive. Tee up the ball, smack it as hard as you can, and measure how far away it landed. Nice and simple, at least if you're on level ground with a "reasonably standard" grass surface. (Half-mile-long gimmick shots on airport runways or icebergs don't count.)

Few golf shots inspire as much awe among fans as a well-hit 1-wood ("Drive for show; putt for dough"), and it's not just about distance. There are a lot of people who can hit 320-yard drives, but if they wind up on the wrong fairway or on somebody's roof, there's not much point to it. That's why PGA Tour spectators only cheer for a long drive if the ball lands on the fairway . . . and on the correct hole. The combination of distance and control is what gets the blood racing.

As it happens, the official drive distances for golfers on tour don't really indicate who hits the farthest, because drive distance is only measured on two of the eighteen holes of the tournament course and the players don't know which ones they are. Generally, though, they're the holes on which players are going to let it all hang out and whack their drivers as hard as they can, rather than the ones with doglegs or lakes or other features that might tempt the players to hit more conservative tee shots in order to protect a score.

Here are the official average drives for the five longest hitters on the PGA Tour as of April 2009:

Bubba Watson	311.6
Robert Garrigus	306.5
Gary Woodland	305.0
Brandt Jobe	303.1
Angel Cabrera	303.0

Notice anybody missing? Legendary big hitter Tiger Woods isn't on the list. Matter of fact, he's not even in the top 100. Woods is capable of outdriving nearly anybody on tour, and at one point loved to pull out the big dog and drive (pun intended) his fans into a frenzy. But somewhere along the line, he figured out that keeping the ball in play was more important than putting on a spectacular but often counterproductive show. One of the things he did when he retreated from the spotlight to reengineer his swing in 2004 was take about 10 percent off his drives in order to increase his accuracy. It worked, and Woods was on his way to becoming the world's first billion-dollar athlete instead of a middle-of-the-pack long hitter.

Bubba Watson's 311-yard average is pretty impressive, but keep in mind that it's an average. The man can hit a ball a lot farther than that, and does so about half the time he tees off. But remember that all of the drives that go into his official tour stats are made in tournaments where his first goal is to keep the ball in play so he can achieve the lowest score possible. Hitting long helps him do that—anything you can do to shorten your second shot is going to increase its accuracy—but hitting long for its own sake is not something tour players care about, at least the smart ones who have a chance at finishing in the money.

So, what about guys who step up and only want to slam the longest drives they possibly can, scores be damned? For a little insight into that, we need look no further than the World Long Drive Championship held every October in Mesquite, Nevada. Entry is not limited to professionals: Anybody who ponies up the $40 fee can take their best shots using a USGA-legal club and ball and try to qualify for the finals. Last year the competition and its $250,000 top prize were won by Canadian Jamie Sadlowski with a record-setting 418-yard drive. (That's less than Jack Hamm's, but under more stringent competitive conditions.)

A monster wallop, to be sure, but how much farther is it possible for a human to hit a golf ball?

•••

L et's start by setting a few ground rules.

It's a badly kept secret that it's possible to make a golf ball your grandmother could hit 300 yards. As a matter of fact, it's a whole lot easier to make that ball than to make one that passes the USGA distance-limiting standard. Here's why:

The USGA has a machine called Iron Byron that hits golf balls with great consistency. It's so precise that the USGA has to resurface the center of its test range periodically because of the damage done by thousands of balls landing in the same spot over and over. (Iron Byron has had seven holes in one.)

Ball manufacturers submit new balls to the USGA for testing. There's a line drawn out on the fairway at a fixed distance from the machine, which hits dozens of balls toward it. If more than five percent of the balls go past the line, they're illegal for sanctioned play.

So the manufacturers are not trying to construct their balls to go as far as possible, which is easy. What they're trying to do is get them to sneak up as close to that line as they can without going over it. They're now to the point where it's just a matter of a few inches. There's nothing left. So when you hear a golf-ball maker tout its new wonder as "the longest ball of all," it's pure hype. You can't make a ball go farther and still have it be legal. All "the longest ball" means is that it "gets closer to the no-no line than the others without going over."

Except . . . you actually can. And if ever there was a testament to American ingenuity driven by the hunger for profits, the Titleist Pro V1x golf ball has to be a contender for the top spot. At least if the rumors are true.

A golf ball's speed depends on how hard it's struck by a club. The harder you hit it, the farther it goes, and that relationship is fairly constant as club head speed increases. What the sneaky engineers at Titleist allegedly figured out is that there is a way to construct a ball so that the relationship is not constant, that beyond a certain threshold, hitting it harder makes it go *a lot* farther. In other words, increasing the club head speed from 110 mph to 120 buys you a much greater increment of speed than jumping from 80 to 90 mph. Exactly how much, assuming it's true, is a deep, dark secret.

What they supposedly did is build a ball that stays linear up to the speed of the club swung automatically by Iron Byron. At that speed, the new ball behaves just like the old ones. But once you get above that speed, it takes off like a rocket. This doesn't do your average player much good, but for the big hitters who can generate high clubhead speeds, the Pro V1x was a revelation. According to the spirit of the USGA guidelines, its use constituted cheating, but there was nothing to be done about it because the ball passed the required tests.

For our purposes, we're going to assume that the ultimate long-ball hitter is using a USGA-legal ball, even if that definition is a little suspect.

If you're a manufacturer, what do you do to find more distance when you can't fiddle with the ball anymore? Obviously, you fiddle with the club. For a long time, clubs were largely unregulated. Then things, as they usually will when left alone and there's money involved, soon began to spin out of control.

A good set of clubs can last a lifetime, which means there's not a lot of money to be made in replacing worn-out clubs. The only thing left to do is come out with "improvements" that induce people to get rid of their obsolete equipment in favor of the new stuff. To do this, equipment manufacturers churn out a seemingly endless supply of new gear, and we spend billions every year buying it. Given that hardly any of it ever makes a difference in how well anybody plays—the national average handicap hasn't budged since the earth cooled, despite all the "improvements"—you have to wonder why people keep buying the stuff. To answer that question, we'd need to get into an extended discussion about marketing and the fluid relationship between advertising and reality, but suffice it to say that if advertising can get people to pay thousands of dollars for tiny diamonds they can't tell apart from glass, selling them golf clubs that aren't much different from their old ones is child's play.

Aside from hyperbolic advertising, the real trick in selling golf equipment is to come out with gimmicks that actually work but that the USGA hasn't outlawed yet, and hope that you can make enough money before they catch up with you. If this sounds familiar, it's because it's a lot like the speedsuits we talked about in the swimming chapter.

One of the gimmicks they came up with was the "trampoline clubface." To understand this, it helps to go through some simple physics, starting with this little mystery: If you hit a golf ball with a club moving at 120 mph, the ball will take off at over 160 mph. How is this possible?

The answer has to do with something called the coefficient of restitution, usually represented by the letter e. In plain English, e is a measure of how much energy is retained when something bounces. If you drop a rubber ball from a height of 10 feet and it bounces back up to 9 feet, e would be .95 (the square root of the 9-foot bounce divided by the 10-foot drop). If the ball were perfectly elastic and bounced back up to 10 feet, it would have an e of 1. If it stayed on the floor like a wad of cake dough and didn't bounce at all, e would be 0. The higher the e, the springier it is. (Just for the record: An e equal to 1 is impossible. A ball that returned to the same height it was dropped from would bounce forever, becoming the world's first perpetual motion machine and the key to solving the world's energy crisis while breaking the most basic laws of the physical universe.)

When a golf ball is struck by a moving club, the club head has its own e, so the ball gets energy not only from the club's movement but from the fact that the club is springy. Springiness occurs when something snaps back to its original shape after it's been deformed. Initially, the club is moving at 120 mph, and most of the force from that impact flattens the ball, while the ball itself hardly moves at all. After that initial impact, in the time it takes the club to travel barely a third of an inch, the ball has already recovered its shape. That happens so fast—three one-thousandths of a second after impact—that the force of the ball popping back into shape pushes it away from the club, and it begins to take off on its own at a faster speed than the club head is moving. A thousandth of a second later, the ball is traveling at 162 mph.

What's harder to see in high-speed photos is that the club face also deforms and bounces back, adding to the speed of the ball. That was the opportunity exploited by club makers. The USGA's Iron Byron didn't test for this in the old days.

Enter the "trampoline" face, so nicknamed because, well . . . it doesn't really need much explaining. It took the USGA watchdogs a while to catch on, but once they did, they implemented new rules designed to prevent balls from being hit so far as to make every golf course in the country too easy to play and therefore obsolete. The regulations dealt directly with e, the coefficient of restitution, and set an upper limit of .83 for clubheads. (Golf balls have an e of about .78). That's going to be our upper limit as well.

It would seem a simple matter to arrive at the longest drive by first figuring out the maximum clubhead speed that a human could generate. After all, the faster the speed, the harder the ball is hit and the faster it comes off the clubface, and speed translates directly into distance. Up to a point, all of that is true. However, the relationship between clubhead speed and ball speed is not as linear as we'd like to think. At some point, the increase in ball speed as a function of clubhead speed begins to dwindle a little. The ball still goes faster the harder it's struck, but the gain in speed isn't as great. The question naturally arises: Is there some threshold clubhead speed above which the ball actually begins to go *slower* as the clubhead speed increases? This might sound strange, but what we're talking about is exceeding the ball's ability to spring back.

Here's an analogy that might help to make it clearer: Imagine that you're on the green putting an apple rather than a golf ball. The harder you hit the apple, the farther it rolls. But when you start taking some serious whacks, you feel the putter face digging into the apple a little instead of bouncing off crisply. Now imagine taking a full swing at the apple with your driver. You're probably going to turn it into apple sauce and splatter the bits and pieces about twenty feet, whereas you were able to roll the apple clear across the 40-yard green by hitting it gently with the putter.

The same thing happens with a golf ball. At some speed, you're going to damage the ball and permanently deform it instead of allowing it to deform relatively gently and spring back into shape. There's only so much punishment golf balls can take, which is why they get mushy over time, like baseballs and Ping-Pong balls.

How does this enter into our calculations? It doesn't. Golf ball manufacturers who have purposely tested their products to destruction have never been able to damage a cleanly struck ball with a golf club, even using machines, so it's a good bet that we're going to reach the human limit well before that becomes a problem. Our quest for the longest drive is indeed a quest for the highest club head speed. Driver swing speed for the average PGA Tour player is about 113 mph. Tiger Woods gets up to 130 mph when he lets it all hang out, and Jamie Sadlowski's head speed when he hit his 418-yard record-breaker was 140 mph. The highest recorded club head speed ever was 163 mph by PGA pro Jason Zuback.

As we said earlier, good technique is critical for a successful long drive. Every fraction of a degree departure from flush means that less energy is going to be devoted to launching the ball toward its target. (*Flush* means that the club face is perfectly perpendicular to the swing path and that the ball is struck with the "sweet spot.") That "lost" energy is used up in imparting sideways spin to the ball, which will cause it to veer to left or right, further reducing the distance.

The ratio of ball speed to clubhead speed is called the "smash factor." The more cleanly the ball is struck, the higher the smash factor, with an upper limit of 1.5. That means that a ball struck perfectly by a club moving at 120 mph will fly off the face at 180 mph. "Perfect" implies only backspin and no sidespin.

While we're on the subject of spin: Back in the beginnings of golf about four hundred years ago, players used smooth balls made of gutta-percha, a hard, inelastic latex produced from the sap of trees in the South Pacific. They noticed that, as the round wore on, the balls flew farther and farther, despite their surfaces getting chopped up after repeated club strikes. Eventually they figured out that it was *because* of the surface damage that the balls flew better. The reason is that the little nicks and scratches bit into the air, creating what we now call a boundary layer, which made the ball more aerodynamic.

A good club strike imparts backspin to the ball. That means that the bottom of the ball is turning toward the wind created by the ball flight. A thin layer of air surrounding the ball that spins with it collides with the air passing underneath, pushing against it and creating lift. It's the same phenomenon that creates a rising fastball in baseball (which doesn't actually rise, just sinks less), and it's the reason all golf balls have dimpled surfaces. Without them, balls wouldn't travel nearly as far.

They also wouldn't hook or slice as markedly, which is what happens when they're hit badly and the spin is sideways. Unless the side spin is intentional. When done on purpose, we call them draws and cuts instead of hooks and slices. Either way, though, it's all by way of saying that technique is critical to the long drive.

Also critical in executing the perfect drive is the launch angle, which is the angle between the ball's flight path and level ground when it first comes off the club face. Optimal launch angles depend on how fast the ball is going to fly off the club. For maximum distance at the kind of speeds we're talking about for the perfection point drive, the ideal launch angle is 16 degrees. How this is achieved is a combination of the club face angle, shaft length, the height of the golfer, and his technique.

What does our ideal golfer look like? Consider what it takes to hit a solid long drive:

1. Muscle power

No getting around this. Golf instructors have a thousand unscientific ways of interpreting the golf swing. All the power comes from the legs; the arms mean nothing. All the power comes from the arms; I can hit a 300-yard drive from my knees so the legs don't matter. Forget the arms and legs; it's the hips. Forget the hips; it's the back. Or the shoulders. The left hand pulling the club down gives you all the power; the right hand means nothing. Or vice versa. Or none of those, because it's the wrists snapping at the last second that do all the work.

Truth is, it's everything. There are twenty-two basic muscles involved in a golf swing, in five major groups: lateral rotators of the hips and of the spine, primary movers of the arms and extensors of the forearms, and primary movers of the wrists. (For the right-handed golfer, the right wrist flexors and left wrist extensors are activated.) The muscles, tendons, and bones all work together in a way that scientists call "kinetic linking." Energy starts in the ground, rises up the legs, courses through the hips and torso, then rushes down the arms and wrists and ultimately into the club. As it passes through each link in the chain, the original energy gets bumped up by the addition of more muscle power. Conditioning of all of those muscles will result in the highest increases and therefore greater clubhead speed, so our long-drive golfer needs to be a strong, muscular guy.

According to the experts we consulted, pure muscle power alone will add 14 percent over Jamie Sadlowski's record drive, an additional 58 yards.

2. LEVERAGE

Take a hatchet eighteen inches long and chop into a log as hard as you can. Now do it with an axe that has the same head as the hatchet but with a handle three feet long. It's going to do a lot more damage to the log. That's because the longer lever arm allows you to channel a lot more power into the swing, which is why you can hit longer golf shots with a long shaft than with a shorter one. (Assuming, of course, that you can still maintain control of the swing.)

But there's another lever involved in the golf swing, and that's your arms. The longer they are, the more power you can generate, so we're looking for a tall golfer to hit our long drive.

This is probably a good time to tell you that world record holder Jamie Sadlowski is only five feet, eleven inches tall, which would seem to render our thesis a little suspect. But it doesn't. If Sadlowski hit the same way he does now but was eight inches taller, he'd hit even farther. Eight percent farther, our experts have calculated, for another 33 yards.

3. FLEXIBILITY

You want as much distance as possible over which to let your muscles work in order to generate maximum power. Leverage is one way of doing that. Swinging around an arc twelve feet in diameter is going to give you more power than a ten-foot arc.

Another way to increase the distance is by using as long an arc as possible. Imagine yourself chopping into that log again but only being able to raise the hatchet two feet into the air. You're going to be chopping a long time. If you're allowed to raise the hatchet over your head, you're going to come down much harder. Now imagine taking it back way behind you and swinging in a big circle over your head. The bigger the arc you swing through, the more power you can keep adding prior to contact.

It's the same thing with golf. Take the club back a few feet before coming down, and you won't get as much power as if you take it back behind you. But in order to get the club back, you need to be flexible. The longest hitters look like Gumby if you watch them in slo-mo. Someone like John Daly, one of the biggest hitters in tour history, has a backswing that's probably about 70 degrees more than the typical golfer's, which allows him to generate tremendous power as his muscles bring the club around. It's not fully additive—you can't apply the same muscle power that far back in the swing as you can in the middle portion—but it still buys us another 4 percent over the current record, or 17 more yards.

4. Technique

One of the most common mistakes that people make when swinging a golf club is attempting to use too much force in striking the ball. While it's true that more force equals greater clubhead speed equals more power, trying to increase that force can actually have the opposite effect. This is because the most important facet of a good golf shot is contacting the ball with the center of the club face and having that club face be perfectly perpendicular to the movement of the club. An overstrenuous swing is likely to mess up both of those things, resulting in loss of both power and directional control, and that's only if you don't miss the ball altogether.

This tells us nothing about what the long-drive hitter needs to look like physically. It just reminds us that, no matter what he looks like, his technique needs to be perfect. Assuming that Sadlowski was just 4 percent off perfect, our experts agree that correcting that would get us 17 more yards.

Add that all up, and our fictional Julian Herrera's drive on the fifteenth at Augusta came to rest at the back of the green, 543 yards from the tee box, the longest drive humanly possible.

The specter of absurdly long hitters has haunted golf for quite a number of years, bringing with it two main worries. The first is that being able to hit too far will render present-day courses obsolete. The second is that the sport is in danger of turning what is supposed to be a test of individual skills into a contest among equipment manufacturers.

Let's take the second issue first. The good news here is that nearly all of the claims thrust upon a gullible golfing community by equipment manufacturers are patently absurd. By "good news" I don't mean that it's wonderful how people waste billions on the golfing equivalent of snake oil; I mean that the sport of golf is in little danger of crumbling due to a plunge in scores because of all of this swell new stuff. I personally have never met anybody whose game improved because of a new piece of equipment. Now, I've met a lot of people who bought new clubs and are hitting *incredibly farther*, putting *amazingly straighter*, nailing greens they could never tackle before, etc., etc. Somehow, though, their handicaps remain the same. This situation is so pervasive that I've actually had serious conversations with club pros who have said to me, with perfectly straight faces, "I'm not saying that your handicap will go down with this new club, but it will definitely improve your game." Since handicap is the best—in fact, *only*—true measure of your game, you have to wonder.

However, there is one claim that manufacturers could make that would be true, and that is that a particular club or ball will let you hit farther. As we said earlier, it's no big deal to make a ball your grandmother could whack 300 yards. The USGA is well aware of that, and has made such balls illegal for sanctioned play. Therefore, the issue of equipment making golf courses obsolete is moot: The guardians of the game are doing a good job in that department.

They're also doing a good job of ensuring that equipment advantages are minimized on the pro tour. While slight or even drastic changes have virtually no effect on the weekend duffer's game, in a tour pro's hands they can make the difference between finishing in the money or slamming his trunk and going home on Friday afternoon because he failed to make the cut. The controversy over the shape of the grooves in a club face is a good example: It sounds like a ridiculously trivial matter, but when millions in winnings and endorsements are on the line, it's not trivial at all, and lack of vigilance can harm the sport.

Or so the commentators would have you believe, which makes this a good point to stop and consider just what something being good or bad for the sport of golf really means.

Let's start with the most obvious example of how this phrase is used. The standard, generally uncontested line is that, at least until a rather spectacular fall from grace, Tiger Woods was "good for the game of golf." Was he?

Millions of people took up the sport or increased the frequency of their playing as a result of the excitement generated by Woods. All of those new players—and their Tiger-wannabe kids—were considered "good for the sport" by every on-air commentator and golf columnist, but what they should have been saying was that Woods was good for the *business* of golf. Equipment manufacturers, course operators, tour promoters, advertising agencies, and sports networks all experienced dizzying boosts in revenues thanks to the millions of fans who enjoy watching golf tournaments and who think that using the same ball Woods uses will make them hit better, but you'd be hard-pressed to make the case that the *game* of golf got better for the millions of people in America who actually play it. I'm one of them, and all I remember about Woods's impact on my experience of the sport was that courses got more crowded, making it harder to get tee times. Nothing about the game of golf improved if you weren't in the golf business.

So in order to take a serious look at things that affect the sport of golf, we have to separate what's good for people who are in the business of golf—including tour professionals, who look as though they're having about as much fun at their jobs as morticians—from what's good for those who play the game for the enjoyment of it.

At the weekend-duffer level, equipment really doesn't make much difference, so long as the USGA continues to strictly limit ball performance and trampoline club faces. Several years ago, I played in a little tournament in which each player was allowed to take only five clubs out onto the course. Just before we teed up, the club president reminded us to be sure to post our scores at the end of the round. I asked him what adjustment factor we should use to compensate for the fact that we could only use five clubs out of the normal complement of fourteen.

He waved it off nonchalantly and said we should just post our scores as we normally would. When I questioned the validity of that, he said, "Trust me: The scores won't be all that much different than they usually are." To my amazement, he was dead right.

And on the pro tour? It's a good bet we can relax there as well. For one thing, drive distance isn't increasing very much anymore. There was a pretty good jump in the late 1990s, but after 2003, average drive distance was increasing by only a yard per year. And even if that wasn't true, it still wouldn't much matter, because the longest hitters rarely do well, and they're getting worse. In the early 1980s, the average rank of the ten longest hitters was 64th

on the money list. Twenty years later, they were down to an average rank of 77th. Their worst year was 2004: 103rd on the list.
What makes this a little odd is that accuracy off the tee isn't as
important on the tour as it used to be. Back in the 1980s, driving
accuracy was closely correlated with money won. Now it barely
seems to matter. So you'd think that the big hitters would have an
advantage; generally, it's tough to be accurate when you're driving
that far but, since the statistics prove that accuracy doesn't matter
much, you might as well go for it and get all the distance you can,
which is a lot for the long guys. Makes perfect sense on paper, but
in reality it just ain't so.

To be sure, there are some mighty big hitters in the top ranks,
but the extra yardage doesn't seem to contribute much to their results. No matter how you slice it (pun intended), pure skill is still
what counts. So unless some method of skirting the intent of the
regulations pops up before the USGA figures out a way to neutralize it, the game as we know it is safe, as are just about all of the
more than thirty-two thousand golf courses in the world.

CHAPTER SIX

HANG TIME

How high can basketball take the dunk?

It's one of the most powerful images in all of sports. Rocketing down court, covering twenty feet of floor with each bounce of the ball, Michael Jordan reaches the top of the key and switches from ground assault to an airborne bombing run. With one last glance at the basket to line himself up, he launches skyward. Flashbulbs pop by the hundreds as he covers the nineteen feet to the front of the rim, the ball held high over his head. Back on Earth, his startled opponents look upward and gape as "His Airness" floats over them. Defending the basket with anything less than a shoulder-fired missile is impossible at this point, and they can only watch, helpless and awestruck, as Jordan finally begins his descent, legs tucked under him like landing gear, and rams the ball through the hoop with such force that it slams into the floor.

Only after it bounces back up and hits him in the leg does he finally return to Earth and touch down.

June 18, 2035, 9:45 P.M.

Four Pennsylvania Plaza, New York, NY

"It's a joke."

The line caught the attention of Dallas Mavericks owner Luke Domini-can. Was this how NBA commissioner Solomon Lax really wanted to sum up three hours of polite, carefully diplomatic discussion? "C'mon, Sol, . . ." he urged as some of the other owners stirred uneasily in their seats.

"Come on, nothing." Lax got to his feet and began pacing behind the table at the front of the room as he ran a hand through his perpetu-ally unruly hair.

"We're talking a lot of highlight reels here," another owner said.

"It's a joke." Lax stopped pacing and began bouncing an imaginary ball. "A guy drives the lane, and if he's within ten feet of the basket—" He lifted a hand high into the air and came up onto one foot—"bada bing: He stuffs it through the hoop while everybody else stands around picking their noses."

"It's a very athletic move," said Bill Carruthers, owner of the Bos-ton Celtics.

"So's ski jumping," Lax shot back. "But how the hell many of those can you watch without it becoming a crashing bore?"

"Crashing's why it isn't a bore," Amos Wilson of the San Francisco Trolleys said with a smile. Then he grew serious and frowned in disap-proval. "I love the dunk. It's one of the most exciting moves in sports."

"You're twenty-seven years old, Amos," Lax replied. "First time you ever saw a basketball game was a week after you bought the Trolleys." It wasn't far from the truth. Wilson was a nerd-turned-financial-genius who'd earned a worldwide reputation as the only person to make money off the Microsoft-Google bankruptcy. "Sit through a full season and you'll be as tired of it as I am."

"But—"

"Sol's right," affirmed Joseph Rooney of the Detroit Pistons, the most senior of the assembled owners. "And so's Bill."

Lax walked to his chair, content to let the emeritus make his point for him.

Rooney leaned back and regarded Wilson as a patient professor would a struggling student. "Dunk's come a long way since the first time a player forced the ball downward instead of just tossing it up and hoping it would drop through the hoop. Guys doing full three-sixty spins, switching the ball from one hand to the other in midflight, facing away from the basket altogether and shoving the ball in backwards over their heads . . ." To the answering nods around the room he added, "And that's just during regulation games. Even the Alley Oop's become just another play." He was referring to what used to be a half-time stunt in which a player catches a pass above the rim and crams it into the hoop.

"But what about the Slam Dunk Contest?" Wilson protested. It was the most popular event at the NBA's annual All-Star Break, a showcase for the wildest and wackiest that players can invent.

"That's different," Lax said. "That's guys doing stuff you couldn't do during a regular game."

"It's like those ice skaters at the Olympics when it's all over," Carruthers said. "That show they put on when the competition's over?"

"I love that show," Dominican said. "I don't even watch the regular competition anymore."

"That's my point!" Lax said. "It's the different stuff you notice! You do the same thing over and over and I don't care how athletic it is, it gets old in a hurry."

Wilson, miffed at having been put in his place by his elders, persisted. "So what do you want to do, Sol? Outlaw the dunk?"

Rooney shook his head. "Not possible."

"Why?" Wilson asked.

"Because it's a legitimate tactical move," Carruthers answered before Rooney could state the obvious. "Once it's started it's hard to defend, and—"

"And, it takes away the risk of missing," Rooney finished for him. "If you can set it up, it's a sure thing."

"Makes a statement, too," Dominican pointed out. Dunking over your opponents' heads was an in-your-face move that added insult to the injury of two points scored.

"Well, there you go," Wilson said smugly, his point having been made for him. "Nothing to be done about it."

Rooney was forced to concede that the upstart youngster was right. "Can't stop it, Sol."

"No, we can't stop it," Lax agreed. He was resting his chin in his hands, tapping the side of his face with one finger. "But maybe we can make it harder. So it doesn't happen as often."

Any number of possible rules changes flashed through the basketball-soaked minds in the room. Only allow a dunk if it's started more than ten feet from the basket? On pass plays only? On a dribble, or from a standing start only?

None of them made any sense. "How?" Rooney asked.

Lax tapped his face a few more times, then said, to no one in particular, "What's the greatest dunk ever?"

"You mean regular season or the All-Star Break?" Carruthers asked. "Ever."

Rooney shrugged, as though the answer was obvious. "Howard."

At the 2009 contest, Dwight "Superman" Howard had the equipment crew crank a portable basket up to twelve feet, two feet over the regulation ten feet, and dunked it anyway, pausing at the apex of his enormous leap to paste a sticker of himself at the very top of the backboard.

Lax, lost in thought, nodded idly and called up his memory of that historic feat. *Make it higher!* Howard had demanded after the spectacular feat. *Higher!* There wasn't any way to do that—the portable basket's height was maxed out—but it did cause a lot of speculation among those present as to just how high a dunk Howard could manage.

Rooney glanced at one or two of his colleagues before looking back at the commissioner. "What the hell, Sol? You saying we should raise the basket?"

Lax waited for a smattering of snickers to die down before he answered by not answering.

"For Pete's sake!" Dominican sputtered when realization dawned.

"Why not, Luke?" Lax rose out of his seat again even as he rose to the occasion. "What's so sacred about ten feet? You think Naismith measured how high up he nailed the peach basket? That's just how high the hayloft was!"

"What about a hunnerd and fifty years of tradition!" Rooney thundered.

"What about it!" Lax thundered back. "Tradition's what's gonna put us in bankruptcy!"

His reference to the impending financial crisis facing the battered league cast the pall he'd intended over the fulminating protests. "We got guys barely five-foot-two slam-dunking the lights out," he said. "Gotten so bad, a guy takes a normal layup, next day the papers want to know why he didn't smash it through the net."

"Still," Carruthers said, "raising the basket?"

The thought overwhelmed them, and there was silence for nearly a full minute, until Amos Wilson, the least steeped in basketball legacy among them, looked up at Lax and asked, "How high?"

Just like that, the conversation shifted from *If* to *How much?*

"I wonder," Rooney mused as he folded his arms across his chest. "What do you suppose is the highest anyone could dunk a basketball?"

Lax nodded approvingly. He'd had the idea of raising the basket but no notion of the method they might use to arrive at an optimal height. Now Rooney had given him a starting point. "Let's figure that out and work backwards from there."

The motto of the modern Olympic Games is *Citius, altius, fortius.* "Faster, higher, stronger." Interestingly, while there's a lot of emphasis on *citius* in the actual contests and a fair amount on *fortius*, there's hardly any on *altius*.

Which is odd, because of all the fantasies people have engaged in since the dawn of civilization, the ability to fly is one of the most persistent and pervasive. The myth of Icarus, who used wings of feather and wax to escape from King Minos but flew too close to the sun and perished in the sea, dates back to over 1200 B.C., and cave drawings depicting human flight predate even that.

People have attempted to take to the skies via any number of contraptions, including hot air balloons, cannons, airplanes, helicopters, jet packs, skis, gliders, trampolines, and rockets. In their imaginations, they've done it on angel wings and magic carpets, and even today they pay money to charlatans promising to teach them the art of levitation.

The quest to fly is endlessly fascinating and compulsive, and new developments tend to captivate the world. It's hard for us in this age of routine airplane flights and shuttle launches to conjure up what it must have been like in an earlier age, but consider that when Charles Lindbergh achieved the first solo flight across the Atlantic in 1927, he instantly became the most famous man on earth and remained so for years. More Americans can name Mercury astronauts from fifty years ago than can name their current congressman.

Since the Wright Brothers' first flight in 1903, which lasted twelve seconds and traveled about the length of a Boeing 737, advances in aviation have come at a dizzying rate, culminating with humankind's first foray to another world in 1969. All of those advances have been entirely technological, since there's very little we can do to climb into the sky unassisted. Stripped naked, we can swim down to remarkable depths on a single breath, and we can run along the ground for many miles on our own two feet, but about the best anyone can do in an upward direction is jump up to reach an apple on a low-hanging branch.

Maybe that's why there's so little emphasis on achieving altitude in the Olympics. We're much better at moving horizontally, so it's the quest for speed that dominates the 388 official medal events of the Winter and Summer Games. Runners cover distances ranging from 100 meters in a sub-ten-second dash to the 26.2 miles of the two-plus hours of the marathon. Rowers, kayakers, swimmers, skiers, speed skaters . . . they're all about getting from here to there as quickly as possible.

Higher is another story. While there are some events in which athletes ascend fairly dramatically from the surface of the earth—ski jumping, gymnastics, freestyle skiing, snowboarding, springboard diving—in none of these is the altitude achieved a significant scoring factor, nor is it even measured.

As it happens, *altius* as the primary objective only applies to two Olympic events: high jumping and pole vaulting. Curiously, though, the actual altitude attained isn't even known. All that's measured is whether the athlete got over the bar. If a high jumper clears a bar set at seven feet, it's considered a seven-foot jump even if there were three inches of daylight between his body and the bar. Similarly, if he brushes against the bar and sets it shaking, as long as it doesn't fall off the supports, that's a seven-foot jump as well.

The high jump is one of the simplest and purest events in all of sports, with hardly any rules other than "There's the bar: Jump over it." It's interesting to watch, especially since the backward "Fosbury Flop" was introduced, but it's hard to get a good feel for how high these athletes are jumping. To fully appreciate it, go to a room in your house with a standard height of eight feet and imagine jumping over the wall if the ceiling weren't there. That's the current world record, and the man who set it, Cuban athlete Javier Sotomayor, is six feet four inches tall, which means he leaped over a bar that was twenty inches above his own head.

It's an amazing feat of strength and agility, yet the sport is so obscure that the only time most people are likely to see it is during the Olympics. It's not hard to see why: How many times can you watch people execute the exact same move over and over, each attempt virtually indistinguishable from the last?

But when it comes to basketball, repeatedly leaving the ground is a spectacle avidly lapped up by millions of fans the world over, even though height is achieved purely by jumping, with no mechanical aids. And while jumping might work reasonably well for fleas and kangaroos, it's not a terribly efficient method of becoming airborne. Michael Jordan's astonishing leaps from the top of the key, endlessly repeated in highlight reels that seem to show him literally defying the law of gravity, resemble "flight" largely in our imaginations, fueled by those replays, the screaming fans, the flashbulbs and, perhaps most compelling of all, the deep desire to fly like a bird and the ache to believe it can be done. But there's a reason those leaps are always replayed in slow-motion: From takeoff to touchdown, the entire performance lasts for all of .93 seconds, tops.

The dismal truth is that no human being can remain in the air for even a single second without relying on something other than his own legs. This may be part of the reason that anyone who can come close to that one-second mark holds us riveted in the attempt.

Which brings us back to the dunk. If a "Who could make the highest dunk?" event was added to the Olympic roster, it would have a good chance of surpassing figure skating and gymnastics in terms of fan popularity.

They call it "hang time," that mesmerizing ability of a basketball player to remain suspended at the peak of his leap on the way to executing a slam dunk. Like home runs, soccer goals, and heavyweight knockouts, it never gets old.

The dunk is real. Hang time, however, is pure illusion. It doesn't exist. As we said, the total amount of time spent in the air by even the highest jumpers in the game is less than a second, and while it looks as if they're not moving vertically at all at some point, it's not true. To understand why, you have to know a few things about the laws of physics, especially the concept of center of gravity.

We're so familiar with gravity that we hardly ever think about it. What you might find surprising is that, of all the forces in the universe, gravity is by far the weakest. To see just how weak it is, simply lift the book you're holding into the air. Do that and you counteract the gravitational pull of the entire planet Earth. On the other hand, if the book were made of iron, a magnet no larger than a toaster could keep a forklift from moving it.

Gravity may be weak, but it's everywhere and it's inexorable. There is simply no way around it, and the laws that govern its effects are immutable. One of them says that, on Earth, an object in free fall is going to accelerate downward at 32 feet per second per second (written as 32 ft/sec^2). That means that, after the first second, it will be moving at 32 feet per second. After two seconds, the speed will be 64 feet per second, and so on, ignoring air friction to keep things simple.

Similarly, an object thrown upward will decelerate at the same rate. If you fire a BB from an air gun straight up at 320 feet per second, roughly 290 mph, it will slow down by 32 feet per second every second. After ten seconds, it will have stopped and will then reverse direction and accelerate downward. (By the way, when it reaches the ground, it will be moving fast enough to pierce a human skull. That's why people can get killed when guns are fired into the air during New Year's celebrations.)

Easy so far. What's a little harder to grasp is that this holds true regardless of what the horizontal motion of the object is. If you drop a bullet from your hand at the same time as you fire one horizontally out of a rifle held parallel to the ground, both bullets will hit the ground at exactly the same time, even though the one that was fired might land half a mile away. If the gun and your hand were both four feet off the ground, for example, each bullet will be in the air for exactly half a second.

This brings us to the concept of center of gravity. Watch a basketball player like Michael Jordan or Lebron James execute a dunk starting nineteen feet from the basket, and you'll see arms and legs folding and unfolding as their knees and elbows bend. Even as they're coming down, their feet might actually be rising, which seems to run counter to what we just said about physics. How do we make sense of all of those disparate movements and relate them to laws of motion first laid down by Sir Isaac Newton and not changed since?

That's where center of gravity, or CG, comes in. CG is a single point within every object where all the mass balances out. It's the point that moves according to the laws of motion no matter what the rest of the object is doing.

In a human being, the CG is roughly behind the navel halfway between the front and back of the body. You've probably seen video of astronauts rolling and tumbling in zero gravity. While the motion looks random and chaotic, it actually isn't. The astronaut's center of gravity isn't moving at all. If you were to paint a dot on the screen where the CG is, you'd discover that it is dead still and that the astronaut is spinning and tumbling *around* it. Unless he pushes off a wall or grabs on to something, his CG cannot move. And if he was already moving across the spacecraft, his CG will move in an absolutely straight line, even if his arms and legs are flailing all over the place.

Now let's go back to our basketball player. If you were to trace the path of his CG as he leaves the floor and flies toward the basket, you'd discover that this point makes a perfect parabolic arc through the air, all the way to touchdown. It doesn't matter if the player is pumping his arms, kicking his legs, tucking his head into his shoulders, or curling up into a ball. Once airborne, there's nothing whatsoever he can do to change the path his center of gravity is going to take. It will decelerate upward at 32 feet per second per second, stop momentarily, and do the same on the way down, while his horizontal speed will remain unchanged (discounting air friction, which will slow both motions down very slightly).

And yet . . . you still can't help thinking that Michael Jordan hangs in the air for a split second longer than the laws of physics call for. You've seen the super slo-mo, and the camera doesn't lie, does it?

As a matter of fact, cameras lie all the time. That's what special effects are all about. But in the case of a Jordan dunk, the camera is actually being quite honest and what you're seeing is an optical illusion. It comes about as a result of how your mind chooses to define "Michael Jordan." Turns out that what you think of as his flight path is really the trajectory his *head* follows, not his center of gravity. And his center of gravity can change in relation to the rest of his body.

If you stand up straight, your CG, as we said, is behind your navel. That means that about half your weight (mass, actually, but let's keep it simple) is above that point and half below it. If you were to lie on your back on a seesaw perfectly balanced so that you were horizontal, your CG would be directly above the pipe that supports the seesaw in the middle. Now imagine bending your knees and pulling your feet toward your bottom. As soon as you started doing that, the seesaw would tilt down on the side your head is on. That's because pulling up your feet moves your center of gravity toward your head. It's no longer directly above the supporting pipe, and so your weight is no longer balanced. To bring the seesaw back to horizontal, you'd have to inch your way down a little so that the pipe was a little higher on your back.

When Jordan takes off, he usually bends his knees and tucks his feet under him. That puts his CG somewhere around his chest. At some point during his flight, however, he extends his legs, which lowers his CG again.

But didn't we say that the path of the center of gravity couldn't be altered once the player leaves the ground? Yes, we did, and it can't. But even though the path of the CG through the air cannot be altered, its location in relation to the rest of the player's body can change. Let's look at what happens when it does.

Picture Jordan in midair with his legs tucked. His CG is somewhere inside his chest. Now imagine him suddenly extending his legs so that his CG drops down to his navel. Or so you would think, but we know that the path of the CG can't be changed in flight. What's actually happening is that his navel *rises* to where the CG already is. In other words, his overall mass is still tracing a perfect arc, but his torso, and therefore his head, are going up while his legs are going down. If your eyes follow his feet, Jordan seems to be falling faster than he should be. But if you follow his head, he appears to float along at constant altitude, as though he were flying (or at least hovering). That's what we call hang time.

Illusion or not, though, physics always wins in the cold war against gravity. His time in the air will be the same no matter what it looks like, and that time will always be less than a second.

In the game of basketball, the amount of time the player seems to hang in the air isn't very relevant. After all, the objective is to push the ball through the hoop in such a manner that the shot is virtually guaranteed to go in. Whether the player took off from the top of the key or from three feet away from the basket doesn't affect the score, even though it certainly affects the crowd and usually the opponent as well.

But we've been talking about a basket mounted ten feet above the floor. Were we to raise it above the standard height, all of that physics starts to become directly relevant. If we could redirect some of the power that goes into moving the player horizontally into getting him up higher in the air, he could dunk through a higher basket.

The question is, how high can the basket be raised and still be dunkable?

What makes a dunk a dunk is that the ball is accelerated downward as it leaves the player's hand. In other words, he doesn't just toss it up and let it fall through the hoop, or drop it gently in. He *throws* it down forcefully. If he does that with especially aggressive vigor, we call it a *slam* dunk. Is there any advantage to dunking? There is, because the player is controlling the trajectory all the way to and down into the hoop. Anything else and he risks being off-center and possibly bouncing the ball off the rim and out of the net.

Slam dunking, however, carries no advantage whatsoever. At least not a physical one. From a psychological point of view, however, the slam dunk is a very clear statement by the player that he's just made an undefendable play, that he's in control, that he's *unstoppable*. If he's been harassed and hounded by the opposition, jamming the ball through the hoop can be an act of vengeance—*Take that!*—or contempt—*You can't stop me!*—or even disdain—*See if I care!* Sometimes it can even be rude: If a team is ahead by twenty points with ten seconds left in the game, there's just no point to sticking it in the other team's face with a slam.

However it happens, though, crowds love it. The Slam Dunk Contest at the NBA's annual All-Star Break is far and away the most popular event of the festivities.

A standard basket is ten feet off the ground measured at the top edge of the thin, circular pipe that holds the net. Many players in the NBA can dunk at that height, as can many college players and some high schoolers. One particular college player in the 1960s, UCLA's Lew Alcindor, was so adept at it, largely because of his unusual height, that the NCAA outlawed it to give teams from other schools a fighting chance. Alcindor would eventually change his name to Kareem Abdul-Jabbar and become one of the greatest scorers in the game's history.

Jabbar was seven feet two inches tall, at least officially. His actual height is long-rumored to have been two or three inches more than that. Whatever the right number, Jabbar could practically touch the rim of the basket just by standing on tiptoe. It didn't take much of a leap to get him up high enough to push the ball down into the basket.

In pro basketball's early days, dunking was frowned upon and rarely seen. Part of the reason was a safety concern, because of the injuries that could result if a player were to be knocked off balance while aloft. Without something to grab on to, there is no way for him to reorient. The other reason had to do with sportsmanship. The dunk was called the "duffer shot" and was considered a selfish and self-centered breach of team-play etiquette.

Then in 1967 came the American Basketball Association. In an attempt to draw fans away from the NBA, the brash ABA, unfettered by what it considered to be antiquated and yawn-inducing conventions of the older league, instituted several innovations. Among them was the three-point line, designed to boost game scores and presumably fan interest. Another was uninhibited encouragement of what has since become the single most thrilling move in basketball, the slam dunk, long disdained by the NBA.

Turning fans on to the dunk proved an easy task for the ABA, thanks to the timely arrival of one Julius "Dr. J" Erving. With his huge hands and dazzling athletic ability, the electrifying Erving handled a basketball like a baseball, putting it where he wanted with uncanny accuracy. Only six-feet-seven but with a vertical leap of forty-one inches, Erving was a human highlight reel who not only elevated the previously disdained dunk to a high art but ensconced it firmly in the minds of basketball fans as a preeminent skill in the game.

In order to dunk, you might think that a player has to have his fingertips pretty close to the top of the ball so he can shove it downward. But if you watch carefully, you'll see that in many cases the player actually palms the ball from the side or even underhanded. That way he can push it horizontally over the rim, then bend his wrist to twist the ball downward, and only give it the final shove when a good chunk of the ball has already descended below the rim. The rim is ten feet above the ground, and the regulation NBA size 7 ball is 9.39 inches in diameter. Assuming that the player palms the ball so that none of his fingers is closer than about three inches to the top of it, we find that, in order to dunk, a player must be capable of attaining a fingertip height of ten feet six inches.

Which brings us to the vertical leap. This is fairly simple in concept. Put a little chalk on your finger, then reach up as high as you can while standing on your tiptoes and make a mark on a wall. That height is called your standing reach. Then jump as high as you can from a stationary position and make another mark. That's called your highest touch, and the difference between the two marks is your vertical leap.

The ability to jump up off the ground without wings is universal among land animals, at least if you don't count elephants. Even snakes can do it, although not very well. The ability is usually described in terms of body height. An adult chinchilla, a breed of rabbit, can jump six feet in the air, which is pretty good in relation to its height of about a foot, but the hands-down winner is the lowly flea, which can jump over a hundred times its body height. That would be like you jumping over a sixty-story building.

The gold medalist for jumping in terms of absolute elevation rather than relative to body height is the puma, a mountain lion that can jump from the ground onto a ledge fifteen feet in the air. By the way, the puma is even more amazing at jumping *down*: The biggest injury-free drop ever recorded for one of these cats was the same as jumping off the top of a six-story building.

Humans are in the middle of the pack when it comes to jumping ability. The vertical leap of the average NBA player is about 28 inches. Michael Jordan's was 43 inches, and Michael Wilson of the Harlem Globetrotters topped out at 55 inches, and Randy Moss of the NFL hit 51.

The world record for pure vertical leap is held by French-Algerian Kadour Ziani, an astounding 60 inches, which is just under a foot shy of his 5' 10" height. Ziani also holds the world record for high kicking, having left a toe mark on a board set just three inches below the height of an NBA regulation basket. Witnessed by television cameras, he once kicked loose a ball stuck between the basket and the backboard. When he dunks, his *head* rises above the rim.

Adrian Wilson of the Arizona Cardinals has cleared a 66-inch bar in a straight jump (that is, with his body always above his feet and landing on his feet), but that's not a true vertical-leap measure because a large component of the altitude his feet attain is the result of bending his legs versus actually rising. To visualize this, imagine yourself standing straight up and then bending your knees suddenly so that your feet come up off the ground. Do this fast enough and you could get your feet twelve or fifteen inches into the air, but your vertical leap would be zero and your head would never rise at all. Similarly, champion high jumper Sotomayor can snake his body over an eight-foot bar, but the technique he uses results in no part of his body getting much more than a foot or so above that.

That's why time in the air, very easy to measure and once proposed as the ultimate in precision, is no longer used as an indirect measure of vertical leap. It's too easy to cheat by simply pulling your feet up just before you land to increase your time off the ground but not your actual vertical leap.

The true measure of vertical leap is fingertip height, or highest touch attained, minus standing reach. But instead of using marks on the wall, NFL scouts use a device called a vertec, consisting of a set of horizontal prongs stacked one above the other half an inch apart and mounted on a pivot. The athlete being tested jumps up from a stationary position and takes a swipe at the prongs. The highest one that swings away on the pivot determines his highest touch. That height is then subtracted from his standing reach to arrive at his vertical leap.

Sounds simple enough but, as is always the case, there's a complication. It relates to how basketball players jump during games as opposed to how they do it when being tested on the vertec.

Highest touch is measured from a squat jump, in which the player stands in one place, bends his knees, and then jumps as high as he can.

But that's not how players jump during games. You'll rarely see a dunk initiated from a stationery position unless the player is trapped by defenders or showing off. Instead, he gets a running start and leaps while on the move.

From our earlier discussion about gravity, we know that horizontal movement has no effect on the rate at which a body ascends or descends. That's why a bullet fired from a gun hits the ground at the same time as one that's simply dropped. And, as it happens, the same is true for a basketball player: Once he leaves the ground, the rate at which his center of gravity rises up and comes down is dictated strictly by the gravitational constant of 32 ft/sec^2.

However, there's a lot the player can do to affect his initial upward velocity. One of those things is to take a step before he leaps.

The reason has to do with a concept called rate of loading, as well as with the stretch-shorten cycle of muscles. When you squeeze a spring and then let go, it suddenly decompresses to its original shape and launches itself into the air. The more you compress the spring, the higher it jumps when you release it.

Something like that happens when you take a leap. If you do it from a standing start, you load the muscles in your legs in somewhat the same way as you load a spring by squeezing it. But if you take a step first and come down harder, you increase the rate of loading, like squeezing the spring more, and more energy is released when you take the jump. That's one of the reasons that high jumpers "jog" to the bar instead of standing still and then leaping. (The other is that they need a little horizontal movement to make sure they come down a few inches beyond the bar instead of on top of it.)

Your ability to take advantage of the step-and-jump is a function of how good your "eccentric loading mechanics" are. For an elite athlete like an NBA player, the step-and-jump adds as much as six inches to his vertec-based vertical leap.

That's an interesting figure: Six inches is what we were subtracting earlier to account for the portion of the ball the player has to carry above the rim. Now we add six inches for the advantage of the step-and-jump. The two cancel each other out, and therefore all we need to know in order to determine how high an individual would be able to dunk a basketball is his highest touch. If we don't have a direct measurement of that, his height and vertical leap will give us a very close approximation.

Although official statistics aren't kept, the highest dunk ever recorded was twelve feet and it's been done twice. The first time was in 2000 by Michael Wilson of the Harlem Globetrotters, who was believed to have a 55-inch vertical leap. The second was the one by Dwight Howard of the Orlando Magic at the 2009 NBA All-Star Slam Dunk Contest.

We don't know the highest touch for these players, but we can estimate them based on what we do know, including a surprisingly accurate rule of thumb for calculating standing reach as 1.36 times height. At the Slam Dunk contest, Howard managed to place a sticker with his picture on it near the top of the backboard *while he was dunking* into a conventional, ten-foot basket. The sticker's height was measured at about 12' 6", so we know he could go a lot higher than that. Given his 6' 11" height, known vertical leap of 40", and calculated standing reach of 9' 5", his presumed highest touch is 12' 9", so he had about 9 inches to spare on his record-tying dunk.

Kadour Ziani has never attempted a twelve-foot dunk, as far as we know, but is it likely he could do it? Just because his five-foot vertical leap is the highest ever recorded doesn't mean that he could dunk the highest basket, because Ziani is only 5' 10" and his standing reach therefore is fairly close to just eight feet. But add that to his vertical leap, and we get a highest touch of about thirteen feet. It's very likely Ziani could dunk twelve feet with a foot to spare.

Which brings us to Michael "Wild Thing" Wilson. He was 6' 4", so his standing reach was close to 8' 7". Combine that with his 55-inch vertical leap, and we get a highest touch of 13' 2", giving him a whopping 14 inches of clearance while setting the world's record.

Impressive, certainly. But what's the highest that a player could *ever* dunk a basketball?

The ability to leap vertically is rooted in the muscles of the legs and hips. While raw power is key, balance, flexibility, and co-ordination are critical as well.

So is technique. Takeoff speed is everything, because once a body is airborne, there are no further opportunities to raise the center of gravity. But while his feet are touching the ground, there are several things the jumper can do to increase that takeoff speed besides the step-and-jump. One is to vigorously swing his arms so that they're on the way up at the same time that the legs are being extended to provide the pushoff. The momentum of the arms shooting into the air will actually pull the body upward.

It's easy to demonstrate if you can find an old-fashioned (non-digital) scale with a needle or dial that indicates your weight. Start with your arms at your sides and then swing them upward with a single, rapid motion. You'll notice that, as your arms are moving up, the needle will show an increase in weight. This is because, when you swing your arms up, you're pushing the rest of your body down. It's as though you were heaving a bag of sand into the air. That upward motion causes an equal and opposite downward motion, according to Newton's third law: For every action there is an equal and opposite reaction.

When your arms reach the top of the swing, the needle will show a decrease in weight as the upward momentum of your arms tries to pull you off the scale. Once you're airborne, though, and no longer in contact with the ground, there is absolutely nothing you can do to alter the trajectory of your center of gravity. This is why you'll see the slam dunker start his jump with his arms down low. As he rises, he brings them up as forcefully as he can without losing the ball, finishing his ascent with ball held high. Not only does that get the ball above the rim, it helps propel him into the air.

Another thing that can affect takeoff speed is the springiness of the player's shoes. "Springiness" refers to a material's ability to take some of the energy going in one direction and return it in the opposite direction. It's why a ball bounces and it also explains how a golf ball struck with a club head speed of 100 mph can take off at a speed of 150 mph.

The same kind of thing happens with a basketball player wearing rubber-soled shoes. As his quad muscles push his feet to the floor, they also cause the rubber soles to compress. As his body rises and the weight is taken off the soles, they spring back to their original shape, pushing down against the floor and up against the player. The resultant vertical leap will be higher than had the player been barefoot. Whatever springiness the soles of the feet have, that of the basketball shoe is additive.

The evolution of the basketball shoe has been a balancing act among comfort and various kinds of performance considerations. Springiness is desirable, for energy conservation as well as jump height, but traction, ankle support, and stability are equally important. Too much focus on any one of those factors can lead to compromises in another.

For our purposes in determining a perfection point for the dunk, we're going to assume an ordinary pair of basketball shoes. Anything else and we could end up with mini–pogo sticks embedded in the soles.

What the perfection point boils down to is understanding the trade-offs between height and vertical leaping ability. Someone with Yao Ming's height of 7' 6" and Kadoum Ziani's vertical leap of 60" is going to be able to stuff a basket over 15 feet high, but is it reasonable to expect that someone that tall could ever get himself that high off the ground? One report out of China has Ming's vertical leap at 26" but knowledgeable observers of the game put it closer to 24". Shaquille O'Neal is 7' 1" and has a 32" vertical. Michael Jordan, who is 6' 6", has a vertical somewhere between 43" and 48", depending on whom you listen to.

Then there's Spud Webb, at 5' 7" one of the shortest NBA play-ers ever. His vertical was about 46", and in 1986 he won the Slam Dunk Contest. Twenty years later, it was won by 5' 9" Nate Rob-inson, who jumped over Webb on his way to the basket. (Webb executed his first dunk the summer before his senior year in high school. He was 4' 11" at the time.)

If we plot the height of a large number of NBA players against their vertical leaping ability, we see that there is a clear clustering and not much correlation in the middle. But it's the outliers at the tall end we're interested in, and that's where we see a clear trend: Once they get above a certain height, their verticals begin to di-minish fairly rapidly. An extra inch of height doesn't buy anything in terms of high dunking if it means two fewer inches of vertical.

According to Dr. Benjamin Domb, whom we met in chapter 3, the height beyond which a diminution in leaping ability begins to compromise the height advantage is 7' 2". Dr. Domb assembled a team of researchers, including Dr. Adam Brooks and Derrick Brown, to answer this question.

"When I first thought about it," Dr. Domb told us, "I would have guessed about six foot eight."

What got him thinking differently? "Dwight Howard," he ex-plained. "I wouldn't have believed that a person that tall could have a forty-inch vertical, but there it is, so we have to believe it."

So does Howard represent the ideal height? "No," Domb be-lieves. "We don't know that he's the highest-jumping man of that height, just the highest who happens to be in the NBA. For all we know, there's someone in a village in Asia who's seven six and can jump fifty inches."

But he doubts it. "From a statistical and physiological perspec-tive," he says, "seven foot two is the limit. Ralph Sampson of the Houston Rockets was seven four and had a vertical of thirty-six inches, which was probably within four inches of the outside limit for someone of that height." (Sampson's career ended after his third knee surgery, but not before engineering one of the greatest moments in NBA playoff history with a buzzer-beating shot that robbed the Los Angeles Lakers of their chance to win a third consecutive championship in 1986.)

After extensive analysis based on biomechanics and muscle physiology, Domb concluded that the highest conceivable stationary vertical for someone 7' 2" would be 51". "That would be a once-in-a-century anomaly," he cautions. "Any attempt to do more than that would result in training injury that would stop the athlete far short of that goal."

How high could this superjumper dunk? "Standing reach is normally calculated at 1.36 times height," Domb says, "but for speculative purposes we would still be within the bounds of reason if we said that our athlete has especially long arms that would let him reach half again his height.

That puts him up to 10' 9" standing on tiptoe, meaning he could dunk without jumping at all. Add in his 51" vertical and we arrive at the perfection point for the dunk, a staggering 14' 5". To put this in perspective, the rim is ten feet above the ground. The top of the backboard is three feet higher than that. The athlete we're talking about could dunk through the *shot clock*.

There would seem to be some pretty profound implications for the sport of basketball. After all, if you have exceptionally tall players who can also jump well, wouldn't they overwhelm the game and make a travesty of meaningful competition? Maybe we could just station a freakishly tall but otherwise unathletic player underneath the basket and let his teammates hit him with high passes all day, which he'd then drop through the hoop with no possibility of being blocked.

That last part isn't something we need worry about: No team is going to leave one man down at the other end and try to defend every opposing possession with just four players. But what about an NBA saturated with freakishly tall leapers? Would it change the game so drastically that it would become a different sport altogether?

As a matter of fact, it would. But it isn't going to happen, for a couple of reasons. The first is that, at least for now, there are no practical ways to churn out eight-foot Goliaths who can vertically leap half their height. Basketball is one of the few sports that favors a very specific physical trait—height—but while you can choose players based on that characteristic, there's not much else you can do to achieve it. You can train to be faster or stronger, but you can't train to be taller. So unless the tallest NBA players start marrying the tallest WNBA players and doing so in droves, it isn't likely that the field of potential giant players is going to get any larger.

The second reason we don't need to worry about the game is that, as we noted earlier, defense is a critical, if underappreciated, part of it. Michael Jordan, arguably the greatest player in the history of the sport, might be best remembered for his near-mystical offensive play, but it's worth reminding ourselves that he was also a nine-time member of the NBA All-Defensive First Team and had a Defensive Player of the Year title to his credit. A team can afford, at most, one powerhouse center with only modest defensive skills—think Kareem Abdul-Jabbar—so populating the club with supertall, one-trick-pony scoring machines just isn't realistic.

The third reason is a little more subtle. We take it for granted that, the taller the player, the better his chances of scoring on any given play because he's closer to the basket than shorter players. It's fairly obvious that taller players have an advantage, especially the center, but it's not entirely about being closer to the basket: A lot of it has to do with being taller than the guys trying to block the shot.

Thinking about Jabbar again, the reason his famous "sky hook" was so effective was that, once he had the ball high over his head, it was virtually impossible to block, because hardly anybody else could reach that high. The opponent's only strategies for shutting Jabbar down were either to keep him from getting too close to the basket to make it a tougher shot or try to prevent any passes from reaching him in the first place.

But when it comes to putting the ball through the hoop, once a player achieves the ability to dunk comfortably, additional height wouldn't help him very much. Underscoring this is the fact that the Annual NBA Slam Dunk contest went twenty-three years before a center finally won it, and centers are traditionally the tallest players. Michael Jordan earned perfect scores seven times during his three contest appearances, a record yet to be broken, and he stood a relatively short 6' 6". That was also Vince Carter's height. Carter mesmerized the crowd at the 2000 Sydney Olympics by performing what would be called *le dunk de la mort*—"the dunk of death"—over a seven-foot French player. Julius Erving, the man who started it all, was only an inch taller than that, and Kobe Bryant is 6' 4" (or 6' 6" or 6' 7", depending on which source you believe). In the main, the best dunkers as listed by *Sports Illustrated* are about four or five inches shorter than typical centers. And during the 2008–2009 NBA season, eight players at least 6' 7" in height never dunked at all (two of them, Steve Novak and Fabricio Oberto, were 6' 10"), while 6' 3" Russell Westbrook dunked 59 times.

In other words, there's no point to loading a team up with big men just to increase the dunk percentage.

But there's another question worth asking, and that is whether the dunk, once considered a rare and spectacular treat for fans, has been cheapened and made less thrilling by overexposure. It's become so commonplace that players feel compelled to add all kinds of showboating adornments to spice things up, because just being able to dunk is no longer considered a sign of special athleticism unless you can do it when you're under six feet tall.

Along the same lines, consider that scoring itself isn't a particularly big deal except in the final minutes and seconds of a close game. In a sport that routinely sees combined scores of well over 200 points, an ordinary field goal barely gets a rise out of the crowd unless the play leading to it was unusual in some way. One of the reasons serious basketball fans find the college game more fun to watch than the professional version is that each point is so much more valuable.

On the other hand, less serious fans love to see point-scoring extravaganzas. This is why in the early 2000s the NBA instituted a series of changes, including revision of the illegal-defense rule, that led to an explosion in scoring along with more dazzling offensive play. The dazzle was good, but the growth in scores? Not so much.

To return a little more excitement to the game and make points more precious, I have a modest proposal: Raise the basket from ten feet to eleven.

Think what this would do for professional basketball. First, the scores would drop precipitously, which would lure back fans who now only tune in for the last few minutes because they feel that nothing that went on before matters.

Second, it would return the dunk to its former glory, because the number of players who could pull it off would be greatly diminished. It would also increase its value: The primary benefit of a dunk is to ensure that the ball goes in. If overall shooting percentage drops because the basket is higher, the dunk becomes far more important in ensuring that points are scored. In other words, the greater level of athletic skill the dunk requires gets paid off not only in fan satisfaction but a higher probability of winning the game.

Raising the basket is not as far-fetched an idea as it sounds. There would be the usual complaints whenever significant rule changes are proposed, starting with the lack of comparability between today's players and yesteryear's. But being able to assess relative athletic prowess by comparing statistics is largely illusory anyway.

Take the most statistics-happy sport in the world: baseball. There are enough numbers racked up in a single game to fill a small book, and new stats are added all the time. The prevailing sentiment is that the game's dedication to numerical analysis directly supports its legacy and traditions so that, as an example, when Roger Maris broke Babe Ruth's thirty-four-year-old single-season home-run record, we could assume they were playing the same game. But they weren't. For one thing, the Bambino whacked his 60 homers in just 151 games. Maris needed 159 to tie the record and 163 to break it. Maris's feat was admirable, but it wasn't as big a deal as Ruth's, and baseball commentators, however badly motivated sometimes, were right to point it out.

Then again, maybe it was an even bigger deal than Ruth's. The baseballs were constructed differently in the two eras, bat manufacturing techniques had changed, officiating was different and, most important of all, they weren't playing on the same ball fields. One of the more remarkable aspects of a sport in which numbers are worshipped almost as much as the players is that the arenas aren't of standard size. Basketball and tennis courts, soccer fields and swimming pools, bowling alleys and Ping-Pong tables are all carefully calibrated to close tolerances. Visiting coaches sometimes carry their own measuring equipment to make sure that everything is legit on the host's territory.

But while the baseball infield may get this kind of scrutiny, when it comes to the outfield boundaries, it's every stadium builder for himself. The distance from home plate to the fences can vary by as much as fifty feet among ballparks. Fenway Park has the shortest fences in the league, and the narrow foul territory adds 5 to 7 points onto batting averages. Can the Red Sox, who play half their games in this hitter's paradise, be accurately compared to players who are saddled with much tougher home fields? And by the way, after Bob Gibson ended the 1968 season with a staggering ERA of 1.12 and the front office started to worry about offense-loving fans abandoning the game, their solution was to lower the mound, thereby shifting some advantage from the pitchers to the batters.

Obviously, raising the hoop a foot is a pretty radical change, but making things tougher might be just what basketball needs. We still wouldn't see the kinds of celebrations that accompany hockey or soccer goals, but scoring a point would become less routine than it is now.

And if we do end up with freakishly tall but still athletic players who make a mockery of an eleven-foot basket?

We'll just raise it another foot.

CHAPTER SEVEN

BREATHING LESSONS

What's the longest someone can hold his breath?

. . . .

At some time or another, each of us has experienced the awful and terrifying feeling of being unable to breathe. Why would someone deliberately go without air for extended periods of time?

Welcome to the strange world of competitive breath holding.

. . . .

Retired United Airlines pilot Bill Graham floated on his back in a pool on the Big Island of Hawaii. After a few minutes, he rolled over, fully submerging his face so he couldn't breathe. He stayed that way as a minute passed, then two minutes. None of the people nearby, one of them right next to him in the pool, others holding stopwatches and clipboards on the deck, seemed concerned.

Graham was nine the first time he jumped into Lake Ontario in his native Toronto with a mask and fins, beginning a lifelong love of being underwater. In his teen years, he and his friends devoured books by oceanographer Jacques Cousteau and underwater motion-picture pioneer Hans Haas, built their own rebreathing equipment, and became skilled spear fishermen. After learning to fly in the Air Force, Graham moved to California at the age of thirty and spearfished all year round, at venues ranging from Acapulco to Vancouver Island. "Filling the freezer was the excuse," he says. "Love of the ocean was the reason."

At the 2:30 mark, he still showed no signs of coming up for air. Three minutes approached . . . still no movement.

Graham wasn't drowning. He was trying to set a new American record for breath holding.

O ur first breath of air marks an instant metamorphosis that redefines us forever, from the submerged marine creatures we've been mimicking in the womb to the air-breathing mammals we've now joined. For the rest of our lives, we will never go more than a minute or two without drawing breath.

Breathing is a strange phenomenon that falls into the nether region between voluntary movements like walking and talking and automatic processes like the beating of our hearts or the digestion of food. It's something that happens entirely on its own but which we can also control. Nobody "forgets to breathe," but we can stop whenever we want, at least for short periods, and we can slow it down and speed it up as well.

We don't usually pay attention to it until something goes wrong. Anyone who has ever taken a CPR class or watched *ER* knows one of the most basic rules of emergency medicine: If someone is deprived of oxygen for more than three minutes, brain damage starts to set in. The longer the deprivation continues, the worse the damage, with brain death a possibility after as little as five minutes.

This is why having your air cut off is such an excruciating experience. It's the body's way of saying, "Listen, you really need to fix this *right now*," and it's why your kid's tactic of threatening to hold his breath until his face turns blue is not one you need to take very seriously, because it simply can't be done.

Yet our typical reaction to the loss of air is a bit odd from an evolutionary point of view. The reasonable response would be to relax in order to conserve the air already in our lungs and think our way out of the situation. Instead, we panic instantly and fight madly to regain breath.

A diver has to be conditioned not to do that. It buys time to switch to a reserve tank, shed heavy gear and swim upward, or signal a buddy to share a mask while they calmly head for the surface together.

All of which makes you wonder: How can a competitive breath holder like Graham go so long without air? And while we're at it, why would anyone even *try* to hold his breath for as long as possible in competition against others who are doing the same?

As you already know, since you're holding this book in your hands, there's not much point in belaboring the second question. There are as many ways to compete as there are people on the planet, because people will find a way to go head-to-head over just about anything. Formerly obscure "sports" such as competitive eating are now covered on television right along with air-guitar tournaments. Curling and rhythmic gymnastics have made it into the Olympic Games, and there is an annual international championship of Rock, Paper, Scissors. Even that dopey and annoying thing your kids entertain themselves with in the back of the car on long trips has evolved into the World Bloody Knuckles Association, complete with rule book, teams, leagues, and world-title events.

And those are just games you've heard of. People are also out there playing petanque, korfball, and crazy pineapple, so it should come as no surprise that they're also competing to see who can go the longest without breathing, an event called "static apnea."

Except, there's that three-minutes-and-your-brain-starts-to-die thing. So what's the point of the competition . . . to see how close you can get to three minutes without air? Is three minutes like a 300 in bowling, a perfect game?

Not quite. The U.S. record for breath holding is over seven minutes. And the man who holds that record is far from brain dead (although his patient and indulgent wife might on occasion take issue with that assessment). So how does he manage not to breathe for over seven minutes when an oxygen-starved brain starts dying after three?

"The trick is not to get your brain to stay alive without oxygen," according to Graham, a student of the sport as well as a top competitor. "The trick is to keep getting oxygen to your brain even though you're not breathing."

At the age of sixty-eight, Graham is an anomaly in static apnea, a sport generally populated by athletes in their thirties. Graham eventually morphed from a casual spear fisherman into a competitive one, and it was at a meeting of his local club that he heard a lecture by world-class free diver and U.S. record-holder Scott Campbell. "He talked about people holding their breath for six minutes," Graham remembers. "By then I could hold my breath for a long time, but I found it hard to believe that anybody could do it for *that* long. But they really could."

He asked around and located people who could teach him the techniques, and he was instantly hooked. "The competitive aspect aside, it allowed me to stay under longer, which is a big advantage in spear fishing." He also discovered the exhilaration of free diving to great depths, and fell in love with that as well.

Breath holders compete by lying facedown in a swimming pool. For spectators, static apnea, which is a branch of free diving, is not quite the same adrenaline-inducing spectacle as, say, basketball or downhill skiing. It's more like watching chess, but without the frenzy. Matter of fact, it might be the only sport in the world in which competitors outnumber spectators, at least if you count the judges and safety monitors.

But as Harry Houdini proved conclusively, the audience doesn't necessarily have to actually see anything to work itself up into a fair level of excitement. The climax of the great magician's stage show was an escape while hanging upside down and handcuffed inside a tank of water hidden behind a screen. Invariably, the audience would draw in its collective breath as the master magician was lowered into the tank, and a series of loud exhalations would then ensue as hundreds of people gave up over the course of the next minute or so. As the clock ticked on and Houdini failed to appear, the audience would begin to fidget, and pretty soon people would start shouting for stagehands to free the showman, who was doubtless drowning because something had gone terribly wrong. Houdini, who had gotten out of the tank within seconds of being lowered into it, would sit behind the screen reading a newspaper, waiting for the exact moment in which to appear, soaking wet and disheveled, to waves of relieved delirium from his audience.

Static apnea competitions are nothing like that. The overwhelming concern is safety, not drama, which is just as well because, unlike Houdini, the competitors are in full view of onlookers, and there's no way to fake it. The contest protocol, at least at first glance, is simple: The competitor lies facedown in water, thereby providing absolute assurance that there's no breathing going on, and a clock runs until he picks his head up again.

Competitions are held in swimming pools, because they're a much safer and more controllable environment than open water.

More important, a calm pool allows the competitor to expend as little physical energy as possible. Energy correlates directly with oxygen consumption, so one of the keys to getting the most out of a lungful of air is not to waste any of it in extraneous muscle movement. Competitive breath holders are capable of muscle relaxation so complete as to make Hindu fakirs envious. About the only muscle in the body doing anything significant is the heart, but over time, a competitor develops some fascinating response mechanisms that reduce even that energy drain.

"As soon as my face touches cool water," Graham says, "my heart rate drops about ten beats per minute." Static breath holders typically get their heart rates down into the 40–45 beats per minute range, whereas the average person's resting heart rate is closer to 60.

There's a wonderfully tongue-twisting phrase in biology: "Ontogeny recapitulates phylogeny." Using humans as an example, it means that an embryo developing in the womb passes through all the evolutionary stages that humankind as a species did, beginning with fishlike gill slits and progressing to such things as a prehensile tail, all of which disappear by birth.

Even though this romantic concept is just a stubbornly persistent myth, there are situations in which dim traces of our ancestry do arise to surprise us. One of those is the ancient response known as the mammalian dive reflex. When we're suddenly submerged in cool water, our pulse slows, a lower amount of blood is pumped through the heart, and blood vessels in our arms and legs constrict, all in an attempt to reduce oxygen consumption and thereby allow us to go longer without breathing.

The dive reflex is built-in; it comes with being human. But there are things we can do on our own to lengthen the amount of time we can go without air. One of them is to increase the capacity of our lungs. All else being equal, the more air we can get into our lungs, the more we have available before the next lungful is required. A typical human has a lung capacity of about six liters, roughly a gallon and a half, but this can be greatly increased through aerobic conditioning.

Lung volume alone doesn't tell the whole story. Only about 80 percent of that capacity is actually exchanged with each breath; the rest is known as residual volume. Competitive breath holders train to reduce residual volume by exhaling as completely as possible and then inhaling as deeply as they can. Over time, the amount of air remaining in the lungs following exhalation diminishes, and more of the lung's capacity is used to bring in fresh air.

A few basics about breathing: When you breathe air into your lungs, oxygen is extracted and infused into your blood. The heart then pumps the oxygen-rich blood out to all the cells of the body. In much the same way that burning wood uses up oxygen and gives off carbon dioxide, cells use up oxygen in the blood and return carbon dioxide to it.

The heart than pumps this oxygen-poor blood back to the lungs, where the carbon dioxide is flushed out and exchanged for fresh oxygen from newly inhaled air. You exhale the carbon dioxide, and the oxygen-rich blood is returned to the heart, where it is again pumped out to the rest of the body. As long as you keep breathing and the heart keeps pumping, the cycle repeats every few seconds for the whole time you're alive.

But what happens if you hold your breath?

At first, not much. Not all of the oxygen you took in on your last breath was absorbed into the blood, so there is still plenty left in your lungs. But every time the heart beats, more oxygen is pulled out of the air in the lungs and more carbon dioxide is returned. After about three minutes, nearly all of the oxygen in the blood has been used up, and because cells can't function without it, their metabolic processes begin to shut down. A lot of the cells in the body can cease functioning for ten or twenty minutes without suffering significant damage. If oxygen is restored within that time frame, they start back up again.

That's not the case with brain cells, which are acutely sensitive to oxygen deprivation. With training, you can boost your lung capacity and reduce your need for oxygen, but one thing you simply cannot train around is the brain's need for oxygen. Without it, brain cells will begin dying after only three minutes. People who climb Mt. Everest without oxygen are able to do so not because their brains can stay conscious with less oxygen than other people need but because their bodies are better at getting available oxygen to their brains. Some of it is training, but a lot is genetic.

So as Graham said, the trick is not to figure out how to keep the brain going without oxygen. It's to keep oxygen going to the brain even if you're not breathing.

One of the ways that competitive breath holders delay the process of brain-cell damage is by building up their lung capacity with strenuous aerobic exercise, in much the same way that endurance athletes such as marathoners and triathletes do. After years of competitive cycling, Graham's expanded lungs can hold more air than an ordinary person's, so he has more oxygen available for his lungs to extract and feed to the blood. The body uses oxygen based on how much it needs, not on how much is available, so the more that's available, the longer you can go without a new gulp of fresh air. If you lose a lung in an accident or one collapses, you can still breathe, although you'll be doing so more rapidly because a single inhalation won't have as much air in it. You experience something similar if you're lying down with something heavy on your chest. Breathing becomes rapid and labored because the lungs can't expand enough to take in a full load of air, so the body compensates by taking more breaths.

However, boosting the amount of air the lungs can take in is only part of the picture, and a relatively small one at that. Graham's lung capacity is only about 20 percent above normal, which isn't enough to explain how he can go without breathing four to ten times longer than other people, even highly conditioned athletes.

Even more important than the amount of oxygen you can cram into your lungs is how efficiently your body uses it. Obviously, the less of it that gets used per minute, the longer the supply lasts. Graham, like all static-apnea competitors, is an expert in full-body relaxation, which reduces the demand for oxygen that powers muscles. But if that were all there was to it, then deep meditators who are in good cardiovascular condition should be able to hold their breaths for five or six minutes, and they can't.

Graham doesn't react to oxygen deprivation the way you and I do. When our air supply is cut off—say, if someone hugs us too hard and too long, or we get trapped underwater—the feeling of suffocation is so terrible that we generally panic, and we'll struggle violently to get the air flowing again. Because of Graham's special conditioning, his response is very different. While it's still painful, it's not as extreme, and he can overcome the panicky urge to breathe. But stoked by competition and absent that urgency, he faces a very real danger, that of blacking out altogether while face-down in water and never breathing again.

While holding your breath for as long as possible seems a relatively uncomplicated endeavor, there's a surprising amount of strategy involved when Graham vies against others doing the same. The first bit of maneuvering involves the order in which the athletes compete or, in the parlance of the sport, "perform." The most desirable spot is last on the list because, as with skiing, weight lifting, and pole vaulting, knowing the best score so far and therefore what you need to do to ensure the win affords a major advantage.

The night before the contest, each participant submits a target "declared" time to officials. Whoever submits the longest declared goes last; whoever submits the shortest goes first, and everyone else is lined up accordingly, so that the longer the time submitted, the farther down the list you go. Clearly, submitting a longer time is what you want to do, except that if you fail to reach it in actual competition, you're penalized. "Picking your declared time," Graham says, "is a tricky balance of trying to obtain an advantageous position while not knocking yourself out of contention completely."

Competitors can dress pretty much any way they want. The goal is to be cool but not cold, because shivering is energy intensive and will completely wreck a performance. If the water temperature is below 80°F, Graham wears a full wet suit of three-millimeter-thick neoprene, including socks. Wristwatch and goggles are optional, as are mask and nose clip.

Graham likes to prep with a series of progressively longer apneas. He'll hold his breath for a minute, then breathe normally for a minute, then hold again for two minutes, take another break, and then do three minutes, all the way to a five-minute hold. This gets his head into competitive mode, and also lowers his heart rate. In addition, these repetitive apneas induce contractions in his spleen, causing it to release its reservoir of red blood cells and enable a longer-duration performance because of the enhanced oxygen-carrying capacity of the blood. This is the same mechanism that allows Weddell seals to stay submerged for over an hour; it's part of the complex, oxygen-conserving dive reflex common to all mammals.

"When I'm finished with that warm-up," Graham says, "my lungs are comfortably stretched, and I've refamiliarized myself with the various sensations I'm going to undergo while performing. By the time the official clock starts, I'm ready, and there shouldn't be any surprises."

Fifteen seconds before his start time, Graham begins doing

something called packing, a trick to force more air into his lungs than is possible with a normal diaphragmatic inhalation.

All air-breathing animals use one of two methods to take in air. Mammals, including humans, use the diaphragm, a large muscle that expands the lungs and brings in air much like a bellows. Other animals, in particular many reptiles, use buccal pumping, in which the tongue or similar organ is used to push air from the mouth down into the lungs. That's the reason for the characteristic repetitive bulging under the mouth of a frog or lizard.

Several decades ago, a technique for mimicking buccal breathing was developed by American free diver Robert Croft. Technically known as "glossopharyngeal insufflation," lung packing overinflates the lungs by using the tongue as a piston to force air in the mouth down through the trachea. An experienced practitioner can achieve about fifteen pumps in ten seconds. A total of thirty to forty such pumps can force well over a liter of extra air into the lungs, causing the chest to expand visibly (and, to novice onlookers, somewhat alarmingly).

Packing, which takes about a year of practice to bring to this level, is not without its downsides. For one thing, overexpanding the lungs is painful. Pain is usually a warning, and in this case it's to warn against rupturing the lung walls or even the diaphragm. The greatly increased intrathoracic pressure can also reduce the volume of blood going to the heart, raising the blood pressure and causing a cascade of negative effects.

When he's finished packing, Graham rolls onto his stomach and two official timers hit the START buttons on their stopwatches.

On November 13, 2003, Graham broke the U.S. record for static apnea at 6:56. A year later, he shattered his own record, holding his breath for 7:39 during a competition in his hometown of Kailua-Kona. And if you search the Internet, you can find video of him unofficially surpassing that record by going 7:51.

Impressive? No doubt. But the current world record for static apnea under official conditions is an astounding 11:35. To get a feel for how long that is, start holding your breath the next time you put a steak on the grill, and don't exhale until it's done medium well. Stéphan Mifsud, the French free diver who set that mark, has held his breath even longer in training, and by the time you read this, there will undoubtedly have been a new official record set.

Graham is reluctant to make a prediction about an ultimate limit. "It doesn't seem that long ago that I thought six minutes was a fairy tale," he says, shaking his head, "and we get smarter about physiology and technique all the time. Who knows what people will be capable of twenty years from now?"

As with any other sport, trying to arrive at the perfection point for static apnea requires that we have a standardized way of assessing results. Fortunately, this odd little corner of the athletic world has a strong governing body, the Association Internationale pour le Développement de l'Apnée (AIDA) and a surprisingly stringent and carefully administered set of rules that ensures comparability among performances. For our purposes, we're only going to consider records achieved according to those rules, one of which is that competitors can only breathe whatever happens to be in the air above the pool at the time of the contest. In other words, they can't breathe pure oxygen out of a tank.

Oxygen in static apnea would be like steroids in cycling or corked bats in baseball: Unless every competitor were allowed their use, the playing field wouldn't be level, and we'd have no way to judge relative performance. Breathing pure oxygen, a trick used by magician David Blaine to remain underwater for extended periods, would change static apnea into a completely different event.

The international rules are designed to ensure fairness, but their primary purpose is safety, because a lot can go wrong when a competitor begins running out of oxygen.

When the breath is held in the lungs and not refreshed, the amount of oxygen dwindles and the concentration of carbon dioxide rises. For the first few seconds, this isn't terribly uncomfortable, because there's enough oxygen for the lungs to absorb and pass to the blood even after quite a few heartbeats. But after a certain number of heartbeats, which varies depending on the individual, the buildup of carbon dioxide starts to trigger alarms in the body, warning us to get a fresh supply of air.

Contrary to popular belief, it's not the lack of oxygen that causes the awful feeling of suffocation. As any pilot who has undergone training in a high-altitude chamber can tell you, hypoxia (literally, "lack of oxygen") resulting from thin air can be surprisingly pleasant. It makes you light-headed and almost aggressively nonchalant, very much like being drunk. At 25,000 feet without an oxygen mask, aviation trainees report that they would be perfectly content to simply sit back and die. Commercial pilots are trained to recognize the early symptoms of hypoxia and do something about it before they happily sit back and not only die but take two or three hundred passengers and an expensive airplane with them.

There's no feeling of suffocation in high-altitude hypoxia because that feeling doesn't arise from a lack of oxygen; it arises from the buildup of carbon dioxide in the lungs. This makes sense from an evolutionary standpoint, because there were a lot of ways for early humans to have their breathing compromised—getting trapped underwater, getting tangled in a vine, getting caught in a bear hug by an enemy or an actual bear—but not a lot of ways for them to accidentally end up at 25,000 feet without an oxygen mask. So the body developed a pretty powerful alarm mechanism for impending suffocation based on carbon dioxide levels but essentially left us with few warnings that the air was getting thin.

Given that, you might wonder how it's possible for a competitive breath holder like Graham to get into trouble. If you or I were lying facedown in a swimming pool, when to come up is not a very complex decision, because feeling your chest about to explode is a pretty good clue.

And the rest of the body isn't exactly staying still. Even before the CO_2 buildup reaches desperation levels, the diaphragm, without its owner's cooperation, will decide it's going to try to draw breath regardless of a closed throat, and will go into spasm to do so. These are significant, painful contractions and Graham might experience as many as a hundred during a contest.

But still, while it might not be pleasant, where's the danger? Pain is a warning, and the athlete can end it by coming up for air. He might lose the competition, but it hardly qualifies as trouble. And yet static apnea competitions are more dangerous than you might think.

At some point the diaphragmatic spasms subside or become less noticeable, and that's about the time that lack of oxygen becomes a much more serious problem than high CO_2 levels. The oxygen deprivation during static apnea causes the same symptoms as high-altitude hypoxia: Graham gets light-headed and disoriented, and his mind begins to play tricks. The apparent passage of time becomes warped, so that seconds might seem like minutes, minutes like seconds, and sometimes both at the same time (whatever that means, which is nothing rational). Like a pilot trusting his instruments in zero visibility, Graham latches onto shoulder taps from his coach every fifteen seconds, as well as time call-outs the coach speaks into his ear along with some encouraging words, to try to maintain some perspective.

More important, the coach is using the responses to those taps to make sure Graham hasn't blacked out or become so disoriented that he's about to. But even if the coach is getting the proper signal back, it's no guarantee that Graham is still rational and the best judge of when to stop. One of the safety policies Graham and his coach agree on when trying to break a record is to quit two seconds after the record is broken, even if Graham feels he's capable of going longer. Another policy his coach insists on is never trying to break a record in competition unless Graham has already done it in practice by a good ten seconds. Competition is harder—butterflies in the stomach rob energy—and there is more pressure to push oneself into dangerous territory.

Blacking out isn't the only risk. There's something jokingly called the samba, but in truth it's not very funny. It's one of the symptoms of primary hypoxia and is caused by low oxygen levels in the muscles. This leads to a loss of motor control, which can look like an epileptic seizure. The diver is said to be "having a samba." In the throes of that violent dance, he can slam his head into the side of the pool or injure his coach or the safety monitor, and he might also be unable to lift his head out of the water. Up until recently, AIDA rules called for the disqualification of anyone exhibiting loss of motor control, however slight. Those rules were amended such that the competitor's time would still be counted if he could hold his head out of the water on his own and complete three tasks, in order, within fifteen seconds: remove his mask or goggles, display an OK signal to the judge, and say "I'm okay" aloud.

Not incidentally, no one has ever died in an AIDA-sanctioned competition, and now you know why we're only going to consider breath holds done under official conditions when trying to calculate the perfection point.

But what about brain damage? What's to stop some overzealous competitor from pushing himself to the point where he starts killing off neurons? Shouldn't we specify in our calculations that the ultimate breath holder stop short of that point?

Turns out there's no need to do that. Not only has there never been a death in sanctioned competition, there's never been any brain damage either, because as long as everybody's paying attention, it's almost impossible.

In order for brain cells to die, they have to be starved of oxygen. The way oxygen gets to them is via the blood circulating in the skull, so as long as that's happening, the brain is in no danger.

But as we've already talked about, oxygen levels in the blood decrease when you hold your breath. What happens if those levels drop below the threshold necessary to keep brain cells alive?

That can't happen, because the heart would have stopped beating long before that. The heart is a muscle that requires oxygen to work. If blood oxygen levels drop below a certain threshold, the heart will stop. That threshold is much higher than what the brain needs to stay alive. So as long as the heart is beating, the brain will get plenty of oxygen. The well-known "three-minute rule of thumb" is actually fairly accurate, but it's also misunderstood: The clock doesn't start when *breathing* stops; it starts when the *heart* stops.

The problem is that the heart's owner may not be conscious while all of that is going on. Blackout occurs around the time that blood oxygen drops to about 85 percent of full capacity. The heart will go on pumping well after that, though, which is why people can fully recover from blackouts, usually within seconds of the oxygen supply being restored. Many pilots who routinely pull high g's during abrupt maneuvers have blacked out dozens of times with no ill effects at all, as have mountaineers scaling tall peaks. (Half the earth's air is below 18,000 feet, which happens to be the altitude at which most people will likely black out without supplemental oxygen.) Blacking out is dangerous, of course, but the danger comes from whatever activity was being engaged in and no longer is, like flying a plane or walking along a cliff or even just standing upright. The danger isn't from lack of oxygen, because there's plenty of cushion below the 85 percent level to keep both the brain and heart functioning without damage.

Static apnea athletes never get past that point. The signaling procedures ensure that no one loses consciousness for more than the fifteen-second interval until the next signal. Competitors might be woozy, disoriented, and convinced they're on another planet, but that happens to Ironman triathletes, bicycle racers, mountain climbers, and people on roller coasters all the time. As long as the static apnea competitor can hold a finger up in the air when the coach taps his shoulder, nothing bad is going to happen to his brain.

The AIDA rules governing competitions give us an ideal framework for determining just how far breath holding can be safely taken. So long as the competitor is in control of his motor functions (no samba!) and can respond to a safety check by raising a finger, the clock keeps running. He must also raise his head from the water on his own when he can no longer stay down, give the OK sign, and say it out loud as well. He will be able to do all of those things if he doesn't black out, and that gives us the first hard number we can use in our calculations.

It's a generally accepted rule of thumb that blackout is going to occur when the blood oxygen level dips to 85 percent of full saturation. We've discussed the fact that there's nothing we can do to get the brain to tolerate less oxygen, but it's also the case that there are differences among people when it comes to exactly what that hypoxia threshold is. Some people can stay conscious enough to respond to a signal at 84 percent, while others will faint dead away at 87 percent.

Our task now is to figure out how long it's possible for someone to go without air before the blood oxygen level dips below the blackout threshold. That's going to come from a combination of the most air we can squeeze into the lungs without causing damage and then using as little of the oxygen in that air as possible.

Endurance athletes like cyclists tend to have large lung capacities. Five-time Tour de France winner Miguel Indurain's was nearly 8 liters, a bit more than Lance Armstrong's 7.5. The highest lung capacity ever officially recorded, using a medical technique called spirometry, was that of Olympic rower and Beijing gold medalist Peter Reed, at 9.38 liters. Some physiologists have estimated that Michael Phelps might have a capacity as high as 12 liters.

That doesn't mean that Reed and Phelps have better aerobic capabilities than Indurain and Armstrong. What matters is lung capacity in relation to body weight, because bigger athletes use up oxygen faster. A Sherman tank holds over 100 gallons of fuel while a Prius holds 11.9, but look at which one gets better mileage and can go farther on a single tank. What we're looking for is the largest ratio of lung capacity to body weight. In other words, a Prius-size car that can hold a Sherman tank–size amount of fuel.

Our ideal breath holder would be a male endurance athlete, tall and thin except for a barrel chest, within which lies a pair of lungs that are outsized for his body. We don't know how big it's possible for a human to get, but because we're more interested in the ratio than the absolute lung capacity, let's assume he weighs 180 pounds with a lung capacity of 13 liters of normally inspired air, and an additional 9 percent after packing, for a total of 14.2 liters. He'd be a nonsmoker raised in a pollution-free environment, because anything that compromises lung function is going to reduce his ability to use oxygen efficiently. He will also have lived at high altitude for extended periods. This will increase the concentration of red blood cells and therefore his blood's capacity to absorb oxygen.

Once we've gotten the most air into the lungs we can, the next trick is to use as little of it as possible. *Use* here has a very specific meaning: What we're talking about is keeping as much in the blood as we can. Oxygen is stored in hemoglobin, the key component of red blood cells, and is drawn off by cells that need it as the blood courses through our bodies. The less that's used by other cells, the more that remains for brain cells.

There are some obvious ways to reduce oxygen use. The easiest to grasp is relaxation. There is a direct correlation between muscle activity and oxygen consumption. The reason you get winded when you run is that your muscles are consuming oxygen at a rapid rate. In order to replenish oxygen in the blood, you have to breathe more rapidly and your heart has to pump faster. The more you can relax those muscles, the less oxygen they'll consume and the longer you can last on a single lungful of air.

Relaxation is key to a long hold, but our ideal breath holder also has to train his legs not to vasodilate when oxygen gets low. Vasodilation, the widening of blood vessels, is the body's way of trying to get more oxygen to the extremities when levels start to drop: If there's less oxygen in the blood, it needs to get more blood to the muscles. Not a bad idea, unless you're trying to get as much blood to the *brain* as possible and your legs don't really need it because, well, you're face-down in a swimming pool. That ability gets built up by repetitively exposing yourself to low-oxygen environments such as high altitude or hypobaric oxygen chambers, or just doing a lot of breath holding.

Another way to keep the legs from vasodilating is to trick them into vasoconstricting instead. The easiest way to do that is to keep them cool, because vasoconstriction is how the body attempts to conserve heat. Narrowing up the vessels exposes less surface area to the cold environment. You're already familiar with this response: It's what causes goose bumps when you get cold. The other effect of vasoconstriction is that blood is directed away from the extremities and toward the heart and brain.

But we don't want the water too cold, because the body's next defense against low temperature is shivering, and that's going to waste a lot of oxygen. Somewhere in the range of 22–26°C (72–79°F) is just about right.

It would seem that it's now a simple matter to determine the ultimate breath hold. We know the maximum amount of air it's possible to squeeze into the ideal athlete's lungs, and we have enough information to calculate the slowest rate at which oxygen can be absorbed. Those two figures should give us maximum time without drawing breath, right?

Not quite. There are some more considerations. For one thing, air and oxygen are not synonymous. Air is mostly nitrogen, which doesn't factor into blood oxygen at all. Only about 21 percent of the air we breathe is oxygen, so only a fifth of our lung volume is usable in terms of cell metabolism.

In addition, air pressure plays a role. The greater the pressure of the air, the more oxygen there is for a given volume. As we said earlier, half the earth's air is below 18,000 feet. That means that there's more than twice as much oxygen available in a lungful of air if you're sipping lattes in Seattle than if you're trekking around Mt. Kilimanjaro's 19,330-foot summit. Therefore, we want our ultimate breath-holding performance to be conducted at as low an altitude as possible. Best place in the world for establishing the perfection point? The Dead Sea in eastern Israel. If you plunked the Empire State Building down on the beach there, the top would still be more than 100 feet below sea level. (That's why it's so dead: There's no place lower for water to flow to, so as it evaporates in the blistering heat of the desert, sediments left behind build up over time. The water in the Dead Sea is one-quarter solid matter.)

The brain requires about 3.3 milliliters (ml) of oxygen per minute per 100 grams of brain tissue. The average adult male brain weighs 1,300–1,400 grams, so assuming that our competitor comes in at the smaller end of the scale, which is what we want, his brain will need 42.9 ml of oxygen per minute. There are about 20 ml of oxygen in every 100 ml of blood, so that's what we'll call 100 percent capacity. Therefore, we can assume a normal correlation that says that a blood oxygen level at 100 percent of capacity delivers 100 percent of the oxygen the brain requires.

At this point you might think that if the blood oxygen level drops to 18 ml per 100 ml, or 90 percent of capacity, the brain will get only 90 percent of the oxygen it needs. Unfortunately for our analysis (but very fortunately for us as humans), it doesn't quite work that way. The brain, although representing only 2 percent of body weight, consumes over 20 percent of available blood oxygen. If the level of oxygen in the brain begins to drop, the body responds by increasing blood flow to the brain. So even though blood oxygen dips below 100 percent capacity, more blood flowing means that the brain still gets what it needs.

But that increased cerebral flow has a limit: Once it doubles, it's going as fast as it safely can. Any more and the increased pressure would cause bleeding into the brain, leading to hemorrhagic stroke. So as blood oxygen drops, we're OK until it hits 50 percent. After that, oxygen levels in the brain will also start to drop.

Brain oxygen is measured as a percentage of complete saturation; 95–100 percent is considered normal. Below 94–95 percent, symptoms of hypoxia, like light-headedness, begin to appear. A saturation level below 86 percent generally results in blackout.

We now have enough information to predict the perfection point of static apnea.

Location: The Jordan Rift Valley
Date: December 2672
    Our ideal athlete is six feet six inches tall and weighs 180 pounds.

He was born and raised in La Rinconada, a mining village of over seven thousand people in southern Peru that has been in existence for nearly fifty years. At over 16,700 feet, only 800 feet shy of North Base Camp at Mt. Everest, it's the highest permanently inhabited town in the world. The athlete, who comes from a long line of high-altitude dwellers, doesn't experience the symptoms of hypoxia until his brain oxygen falls below 93 percent. He won't black out until it drops below 85 percent.

His lung capacity is 13 liters. With packing, he can squeeze in another 1.2 liters without risking a potentially lethal lung injury. Because of residual volume, only 90 percent of it is usable, and only 21 percent is oxygen, so the total available oxygen in his lungs is 2.7 liters.

Someone living at an altitude near sea level has a blood oxygen-carrying capacity of about 20 ml of oxygen per 100 ml of blood (20 ml/100 ml). But because of the greatly boosted red cell density in our ideal athlete's thickened blood, there is enough hemoglobin to carry 35 ml/100 ml. When at home in La Rinconada, he never gets even close to full capacity, but he will during his static apnea performance: It's going to be held at the Dead Sea, just east of Ein Gedi, Israel. At nearly 18,000 feet lower than his home elevation, the partial pressure of oxygen will be twice what it is at La Rinconada. For him it will be like breathing in a hyperbaric chamber, but in full compliance with AIDA rules because he's taking in only normal, ambient air.

The competition is held in a pool on the beach because the waters of the Dead Sea are not the kind you want to put your face in. For the first three minutes of the hold, oxygen will be absorbed from the athlete's lungs into his blood at the rate of 1200 ml per minute. It's far more than he needs to keep his brain O_2 levels at 100 percent but there's no way for him to slow it down.

After three minutes, the absorption rate begins to drop. Sometime in the sixth minute it's at only 200 ml/min and, at that rate, 100 percent brain oxygen saturation can no longer be maintained. Even as an uncomfortable pressure in his chest begins to gnaw at him, a survival mechanism triggers his heart to beat faster and get more blood flowing to the brain. This doesn't put more oxygen into the blood; it just gets more blood to the brain.

The blood flow continues to increase, but after another four minutes, it's hit twice it's normal rate and can't go any faster. Oxygen absorption from the lungs into the blood is down to 50 ml/min, and his brain starts taking emergency measures to keep itself alive. Blood flow to the extremities decreases in an effort to shunt more to the brain.

Despite these measures, oxygen saturation in the brain starts dropping. There simply isn't enough O_2 available to keep the level at 100 percent. In another two minutes, it's down to 93 percent, and the athlete begins experiencing the symptoms of hypoxia.

He feels himself swaying, rotating, as though swept by a current. The touch of a hand on his shoulder momentarily orients him and he realizes he's perfectly still, not moving at all. A small voice arrives, as if from a great distance, and tells him the elapsed time. He raises a finger in the air to let the coach know he's alright, and that small act grounds him once again.

But it's short-lived. Most of the air in his lungs has now turned into CO_2 and his body, in a vain attempt to get him to seek fresh air, is assaulting his chest. The pressure has turned into pain; the gnawing has taken on the hint of suffocation.

The flow of oxygen from his lungs into his blood continues to slow, and the levels in his brain drop even further. After several more 15-second signals and responses, he begins hallucinating, his inner vision alternately brightening and darkening. He barely understands the time calls anymore, and his signals back are largely based on reflex. Shadows swim before his closed eyes and he has to fight to remember who he is and why he's not breathing.

Something lands on his shoulder; almost too late he remembers what it is and raises a finger. The hesitation puts his coach on alert. Fifteen seconds later it happens again, but again the athlete got the finger up in time. Or did he? He raises it again, just in case, and now the coach is worried: Is he really responding purposefully or is he just on autopilot? He's going to shorten the interval to ten seconds.

The athlete's lung-to-blood absorption rate is down to 8 ml/min and his brain O_2 level is at 86 percent. His chest is screaming, and there's a volcanic eruption inside trying to blow him apart. Poised at the very edge of blacking out, he feels a touch on his shoulder but it's more than a touch this time; it's a pain. The coach is squeezing him, hard, and he won't let go.

With everything left of his mental faculties the athlete forces himself to concentrate. He opens his mouth and exhales, a cloud of carbon dioxide bubbles boiling to the surface. He keeps his head in the water—no sense surfacing and stopping the clock if all he's going to do is exhale—and when the last of the gas is expelled he raises his head, tears off his goggles and nose clip, makes a circle of his thumb and forefinger and says "I'm OK!" The judge acknowledges with a nod, and that makes the time official.

The athlete gasps, inhaling mightily. The putrid air of the Dead Sea is like the purest nectar he's ever tasted.

It's the first breath he's drawn in fourteen minutes and forty-seven seconds.

That's the perfection point for a "normal" human being. But what about someone who isn't quite normal and also wasn't restricted to AIDA competitive rules?

We decided to investigate how long a human could hold his breath if he were allowed to breathe pure oxygen first, and also if he was somewhat abnormal physically but well within the bounds of possibility.

He would be a male seven feet four inches tall weighing 202 pounds. He would have a Marfanoid body type with long torso, long fingers, short legs, and lungs capable of holding 16 liters of air. Marfan's syndrome, the condition he suffers from, was also thought to have affected Abraham Lincoln and the violinist Paganini.

He'd have been born and raised at base camp on the south side of Mt. Everest in Nepal. At his home's elevation of 17,700 feet, most casual visitors would pass out from lack of oxygen, but over the course of his life, his body adapted by using more of his lung capacity to compensate: His residual volume is only 4 percent. His blood hematocrit is 56 percent, indicating a level of red blood cell concentration that is just shy of causing clots that would kill him, a condition called polycythemia. His blackout threshold is 82 percent brain oxygen saturation.

If this unusual human breathed pure oxygen under high pressure in a hyperbaric chamber, then packed his lungs before going face down to begin his static apnea performance, he would not reach severely hypoxic blood oxygen saturation levels for approximately 32 minutes.

CHAPTER EIGHT

THE LAST SLUGGER

Hitting the longest homer

Baseball is ninety percent mental.
The other half is physical.
—Yogi Berra

You can talk all you want about pitchers' duels and suicide squeezes and balletlike double-play throws from second to first, but the unassailable fact remains: Nothing gets a baseball fan's blood pumping like a moon shot of a wallop that sends the ball right out of the field and into the stands or the light fixtures or the street or, in the case of the Giants, the waters of San Francisco Bay.

The home run is literally a show stopper. In almost every other situation, players are running after balls or running after base runners or shifting positions in anticipation of how the play will develop. "The game is afoot," as Sherlock Holmes liked to put it, and the field is alive with movement and tension.

Not when there's a homer. As soon as everybody realizes that the ball is riding a one-way ticket to bleacherville, everything pretty much comes to a dead halt. Fielders watch helplessly as the unplayable missile sails its way into oblivion, the pitcher hangs his head in sullen disgust with himself, and the only things moving on the field are the hitter and any of his teammates who happened to be on base. They're not so much running as loping, in no particular hurry to get anywhere, their trip around the bases being purely a formality, logically unnecessary but required by the rules.

At least that's the way it is on the field. In the stands, it's another story. A home run is an automatic standing ovation as inevitable as a Pavarotti curtain call. If it happens in a late inning with men on base and the heretofore lagging home team suddenly thrust into the lead, the frenzy will be especially fevered but, even if it's just one more run tacked onto a ten-zip wipeout, the crowd will burst into a raucous, lusty cheer as it jumps to its collective feet. Every single time. Or almost every time.

The people running baseball know that home runs put butts in the seats, and they've done everything short of Flubberizing the ball to give the fans what they want, including looking the other way when 180-pound journeyman hitters began showing up as 220-pounders, smacking thirty and forty homers a season when they used to hit eight or nine. Rumors that the balls were juiced to make them fly farther have largely dissipated. Rumors that the *players* were juiced have, sadly, been repeatedly and distressingly substantiated.

It's interesting that, unlike in any other sport, baseball has a small but vocal contingent claiming that records put up by sluggers subsequently revealed as steroid users ought not to be accompanied by the dreaded asterisk denoting "achieved under questionable circumstances." The rationale is that performance-enhancing drugs have nothing to do with achieving the hardest feat in sports: taking a round bat to a round ball and hitting it square. That may be true, but hitting it square 400 feet instead of 350 is another story. If it weren't, why were those record-breaking home-run hitters shooting D-bol and Anavar in the first place?

The home run has become commonplace. The fans still stand, but the intensity of the celebrations has diminished except when the homer is of significance in the outcome of the game. After all, how can you expect people to stomp and shout with every bomb when there's an average of over one homer per game throughout the league?

Much of the interest now has shifted from "Who smacked a homer?" to "How far did he hit it?" Sneaking one over the fence is journeyman stuff. Popping one into the middle deck is exciting. Slamming a towering shot onto the roof of the redbrick Western Metals Supply building in left field at San Diego's Petco Park? Now *that's* thrilling.

Which leads us to wonder: What's the longest home run it's possible for a human to hit?

The home run ain't what it used to be. As it matter of fact, it didn't used to be anything. In the early days of baseball, fans stood behind ropes in the outfield and any ball hit into them was a ground rule double. Homers, which didn't even have a name back then, were a rarity.

As late as 1883, there still weren't enough of them being hit for the feat to be considered a "standard" aspect of the game. On opening day of that year, Roger Connor of the New York Giants belted a long one at the Polo Grounds that was talked about in hushed awe for the rest of the season . . . which probably had to do with the fact that it was the only one he hit all season.

Because the mere act of hitting a homer was an event in itself, nobody back then was very serious about accurately measuring the distance. Think of it like the early days of parachuting: The big deal was to hit the ground without dying, and only after jumping out of airplanes became routine did people get around to attempting pinpoint landings and thirty-man formations. Similarly, it was only after the home run became more conventional that serious thought was given to accurately measuring the distance. Early reports of 500-foot home runs are about as trustworthy as estimates of crowd size at protest marches.

What turned the home run mainstream and thereby spurred more reliable distance measuring was the arrival of a kid named George "Babe" Ruth, the Sultan of Swat, who turned the home run from an occasional oddity into the premier attraction of the game. If you're tempted to grudgingly grant Ruth credit as the first in a long line of home-run kings but dismiss him as having been easily surpassed by his modern counterparts, consider this: In his rookie season with the Boston Red Sox, the Bambino smashed one out of Sportsmans Park in St. Louis, over the right field bleachers and across Grand Boulevard, hitting the sidewalk a carefully measured 470 feet from home plate. In the thirteen years from 1982, when IBM systems for measuring hitting distance were first installed, until 1995 when every major league ballpark had one, there was only one home run that went more than 500 feet. Even Jose Canseco's legendary blast into the fifth tier of Toronto's Sky-Dome in 1989 was only fourteen feet longer than Ruth's measured shot.

The new measuring technology, which took into account the physical realities of a ball's trajectory, destroyed a lot of legends. Home runs that had been routinely called 500-footers if they hit some fixed stadium structure were reevaluated as 450-footers under the cold scrutiny of unemotional electronic instruments.

Ruth himself almost certainly hit homers over 500 feet but,

in those days, unless the ball actually left the park and hit level ground, there was no good way to measure the "assumed" distance of a ball hit into the stands instead of returning to the same elevation as the ball field. Inferring the distance is a bit tricky: If the ball traveled 250 feet before reaching its apex, it can't be assumed that it went another 250 before touching down again, because the path of travel isn't symmetrical except in a vacuum. Owing to friction with the air, most of a long ball's forward motion has dissipated by the time it nears the ground, and it drops almost straight down at the end. To give you an idea of how significant air friction is, recall that a pitched ball loses about one mile per hour of speed for every seven feet it travels on its way to the plate. It may have been a 100-mph fastball when it left the pitcher's hand, but it will only be doing about 93 mph by the time it reaches the catcher's glove.

In order to predict the longest possible homer a human could ever hit, we have to know what kind of pitch produces the longest ball. Simple physics tells us that we want the fastest pitch possible, because the ball's forward energy will be returned when it bounces off the face of the bat. To prove this to yourself, imagine throwing a ball against a wall: The harder you throw it, the faster it's going to bounce back at you.

That would seem to complicate things a little, because in order to figure out the longest possible homer, we have to figure out the fastest possible pitch. The good news here is that we don't need to worry too much about that because, unlike in many athletic events, pitching doesn't seem to be getting any better, or at least any faster. In fact, there aren't any more truly fearsome fastball pitchers today than at other times in baseball's storied history. The fastest pitch ever officially recorded was a 104.8 mph meteor by Detroit's Joel "Zoom Zoom" Zumaya on October 10, 2006, against Oakland at McAfee Coliseum, and it's doubtful it can get much faster. At least by a conventional pitcher.

It's actually possible to throw harder than 105, and some pitchers probably already have. The problem is, it may not be possible to do that and still have enough control over the ball to put it over the plate. A pitcher not only has to generate power, he has to "stay within the pitch" to avoid throwing it at the mascot or into the bleachers, in the same way that a golfer can't "come out of the shot" by applying so much power he loses control over the swing and misses the ball completely. The limit for a big league pitcher's ability to combine power with control seems to be 105.

At least according to our notion of a "conventionally useful" pitcher. Those who can throw only one kind of pitch don't generally make it into the Bigs, because their predictability is a fatal liability. (An exception is Tim Wakefield of the Boston Red Sox, who throws knuckleballs and is one of the very few pitchers ever to have struck out four batters in a single inning. In his case, the unpredictability is built into the pitch itself, since even Wakefield doesn't know what the ball is going to do on its way to the plate. I'll explain a little later.) That would be especially true of a pure fastball pitcher, because a fastball has less "stuff" on it than other kinds of pitches, the primary advantage being that it gives the hitter less time to assess its trajectory. But what if there was a one-trick pony fastball pitcher so powerful that the speed of the ball was enough to foil hitters even if they knew what was coming because there would be less time to figure out the path of the ball? Assuming that the pitcher could achieve enough control because he only had one pitch to deal with, how fast would it be conceivable for him to throw and still get it over the plate half the time?

According to Dr. Bassil Aish, the limiting factor is not muscle power or technique. It's how hard he could throw without dislocating his shoulder or tearing a rotator cuff or pulling a tendon off a bone. Aish has calculated that an appropriately proportioned pitcher could get strong enough to throw well above 105 mph but that anything over 111 is almost sure to result in serious damage. And in case you're thinking that he could still get off a single 115 mph pitch even though it would be his last, Aish says otherwise. "The damage would occur partway through the pitch," he discovered, "and he'd be lucky not to throw the ball into the dirt."

If you think serious injury as a result of throwing a hard pitch is just theoretical, just ask Tony Saunders of the Tampa Bay Devil Rays, who broke his humerus, the bone that runs from the shoulder to the elbow, on a 3–2 pitch to Juan Gonzalez of the Texas Rangers in 1999, or San Francisco's Dave Dravecky, who broke the same bone while throwing a pitch ten years earlier. Cleveland's John Smiley broke his left humerus while warming up for a game in 1997, and Cincinnati's Tom Browning did the same while throwing a pitch in a game in 1994.

None of those guys was pitching anywhere near 111 mph, which is what we're going to use for our perfection point pitch.

It seems simple now: Just launch a 111 mph missile and let the batter hit it with everything he's got at an optimum launch angle. But it isn't that simple, because there's air friction involved. That's going to slow the ball down, and while there's not a lot we can do about that, there is a little. As long as we're going to encounter friction, we might as well use it to our advantage. We do that by imparting some backspin to the ball when it's hit. One of the reasons a lot of hitters like the "hanging curve" (a mistake pitch that doesn't curve as much as the pitcher would have liked) better than the fastball is that the curve provides a better chance of putting backspin on the ball when it's struck. Increasing the chance is important because there's no way to purposely hit the ball just below center; it's tough enough to hit it at all.

Backspin means the bottom of the ball is spinning toward the oncoming air as the ball takes off. This increases the air pressure underneath the ball, making it rise at a higher rate than it would have just based on the launch angle. If you've ever hit a golf ball at a very low angle and then watched it actually climb more steeply as it flew away, you've seen this phenomenon, which is also the secret behind a rising fastball. (A misnomer, actually: It doesn't so much rise as just sink less.) The fact that a baseball has low density, meaning that its weight is low for its size, increases the aerodynamic effect. If the ball were made of lead, spin would hardly affect it at all because there's so much more mass for the air to try to push.

It's all about spin, which is how a pitcher puts "stuff" on the ball: By spinning the ball in different directions as he releases it, the pitcher can throw a slider, a curveball, a cutter, or, if he manages to throw it with no spin at all, a knuckleball. The knuckler is a very slow pitch that ought to be easy to hit but, because the ball isn't spinning, the stitches holding the cover together affect the air randomly, causing the ball to dance crazily on its way to the plate. Not only don't the pitchers or batters know what the ball is going to do, neither do the catchers, many of whom use special oversize mitts to keep the ball from getting away from them.

That anyone can even hit a big league pitch is a wonder in itself. That some can hit home runs is practically a miracle. On paper, at least, the feat seems impossible.

The mound is 60 feet 6 inches from home plate. But by the time the pitcher releases the ball, only his heel is touching the rubber, so he's about five feet closer to the plate. If he throws a 99 mph fastball, the ball is going to reach the batter in less than four tenths of a second, 395 milliseconds (ms). By comparison, it takes 400 ms—four tenths of a second—to blink your eye completely.

A lot has to happen in those 400 milliseconds. It takes the first 100 for the batter to see the ball in free flight and get an image to his brain. The brain then needs 75 ms to process the information and gauge the location and speed of the ball. In the next 25 ms—a fortieth of a second—he has to decide whether to swing, and then he's got only 25 ms more to decide if the ball is going to be high or low, inside or outside. If the decision was made to swing, another 25 ms are needed for the legs to react and begin the first motions of the swing. That leaves a grand total of 150 milliseconds for the hitter to get the bat around and make contact.

And those are only for the gross movements involved. There's still some fine tuning to do. If the batter is only 7 ms early or late in connecting with the ball, he's going to send it foul. And even if his timing is perfect, he still has to put the "sweet spot" of the bat within an eighth of an inch of the correct spot on the ball. To give you an idea of the margin of error, the width of an average pencil is *twice* as big as the margin of error on a major league bat. Making the task even more difficult is the fact that the ball is literally invisible to the batter when it's within the final fifteen feet in front of the plate. The ball is traveling so fast that the brain simply can't process the image quickly enough. No batter has ever actually seen a major league fastball hit the bat. His brain only thinks it did because it extrapolated the remaining trajectory. This explains why even great hitters completely miss so many pitches: If the ball's actual path over the last fifteen feet doesn't match their mental extrapolation, the ball isn't going to end up where they think it will.

The ball, meanwhile, isn't cooperating. The pitcher has put spin on it, making it curve or drop or both, and the batter has to anticipate all of that and make sure that the bat is where the ball is at the moment it crosses the plate. To top it off, he has to swing pretty hard. If he's going to hit a home run, he has to swing *very* hard and as every golfer, tennis player, marble shooter, and place kicker knows, the harder you try to hit, the tougher it is to hit with accuracy.

If you told all of this to an alien freshly landed from Mars, he'd refuse to believe that anyone has ever hit a home run, except maybe by pure luck once every twenty-five years or so. And he's never going to believe that Reggie Jackson smacked one into a light pole on the roof of Detroit's Tiger Stadium. Maybe this explains why even a once-in-a-generation athlete like Michael Jordan wasn't able to cut it as a baseball player, and it also explains why a .300 hitter is such a prized member of a baseball team. In what other human endeavor does failing 70 percent of the time make you a star?

Yogi was right: Baseball is 90 percent mental. The other half is physical.

The easiest pitches to hit are the ones that are the most predictable. During the home-run derby at baseball's annual All Star Break, hitters routinely belt balls into the stands one after the other. In 2008, Prince Fielder hit 23 homers to win the contest. The reason for the seeming ease with which homers are hit at the Derby is that the pitches are coming from the hitter's own batting practice pitcher, serving them up exactly the way the hitter ordered them. The batter knows exactly what the ball is going to do and doesn't need to spend a lot of mental energy deciding whether to hit or not. He just needs to watch it come in and gauge when to start his swing. By the way, the pitchers aren't serving up blazing fastballs during the derby. Even though they'd result in longer balls, there's less time to zero in on their exact position as they approach the plate.

But we're not going to concern ourselves with how easy it is to hit the perfect home run: After all, our guy only has to hit it once. All we care about is the type and speed of the pitch when it does get hit, and the pitch we want is a fastball. Optimally hit, a curve would result in a longer flight than any other kind of pitch because it provides the best combination of velocity off the bat and backspin. "Optimally hit" means that the sweet spot of the bat connects just below the center of the ball, supplying the necessary backspin, while the batter is swinging upward at an angle of 35 degrees, which gives us the ideal launch trajectory.

However, it isn't possible to throw a curve as fast as a fastball, and the difference in speed more than makes up for the advantage of the curve ball's other characteristics.

There are other factors that come into play. For one thing, we want the driest ball park possible, because the "springiness" of the ball decreases as the air gets more humid, resulting in less distance. While it's true that, for a given speed off the bat, a baseball will travel farther in humid air than dry air (counterintuitive but true: humid air is actually less dense than dry air), a dry ball hit on a dry day will travel farther than a moisture-laden ball on a humid day.

Temperature counts, too. Cold air is denser and puts more drag on the ball. When the air temperature is 75 degrees or higher, about 4 percent of batted balls result in homers. When it's colder, that number can drop as low as 3.2 percent. To decrease air density even further, we want a stadium at high elevation.

Wind speed enters into it as well. As we saw with pitching, a baseball is an aerodynamic object because of its low density and the stitches sticking into the air. Even a light breeze eddying around in an open-ended stadium can mean the difference between a home run or a caught fly ball. Backed by a 30 or 40 mph tailwind, a well-hit ball hanging in the air for six or seven seconds could gain an extra 60 to 90 feet of distance.

For our longest-ball prediction, however, we're going to assume a windless day. Picking the highest possible wind would be purely arbitrary because, whatever we chose, we could always add more speed or throw in a rare but plausible full-force gale blowing in exactly the right direction. More important, we're trying to calculate the ultimate human performance, under repeatable conditions, eliminating any extraneous factors that the athlete can't control. A windless day will give us a solid base, and a simple adjustment can provide variations on the perfection point homer for any wind condition we care to dial in.

All we need now is the fastest possible bat speed.

One question we need to deal with is whether it's possible to hit a baseball so hard it gets damaged and therefore doesn't fly as far.

The answer is a resounding *No*.

The *Mythbusters* television show did a segment during which they tried to see if they could literally knock the cover off a baseball, using an air cannon to simulate batting the ball. They weren't able to cause any damage until they cranked it up to 435 mph, more than four times the speed any human batter has ever swung, a remarkable testament to the resiliency of a baseball. (Makes you kind of sad at the waste, too: During a typical baseball game around seventy balls are used. Most go into the stands as foul balls, to be caught by souvenir-seeking fans, never to be thrown or hit again.)

We're now free to swing away as hard as humanly possible, and that's where things get a little complicated. The easiest way to talk about what we're looking for is to start with someone we already know who has one of the highest measured bat speeds in the game, and that would be Derek Bell.

Bell established himself as a power hitter early in his career when, still in the minors, his .344 average earned him the batting title of the New York–Penn League. He went on to win the International League Most Valuable Player award, and *Baseball America* magazine named him the Minor League Player of the Year. His best year as a pro came in 1995 with the Houston Astros, where his .334 average was the fourth highest in the league.

Bell was six two at 215 pounds in his playing days, and his highest recorded bat speed was an incredible 95 miles per hour. Using this as a starting point, we'll break the swing into four major components to simplify things and then take it from there. We'll assume a standard, rules-conforming bat made of maple, which is fast replacing ash as the wood of choice for long-ball hitters because it's denser.

1. Upper extremity rotational speed

According to Dr. Aish, a batter's hands, wrists, elbows, and shoulders form a "kinetic chain" enabling the linked sequence of motions that results in rotational acceleration of the upper extremities. At the "extreme of the extremities" there's a baseball bat, an artificial, highly leveraged extension of the body. The purpose of all that rotating is to get the sweet spot moving through the air as fast as possible.

There are two ways in which our ideal hitter could increase bat speed resulting from rotational speed. The first is simply by having longer arms. Swing a rock tied to a one-foot string around your head at one rotation per second (60 rpm), and the rock will move at about four mph. Make it a two-foot string, and the same rotational speed will get the rock moving twice as fast. Similarly, for any given rpm, the farther you can get the bat from the center of rotation, which is the center of the batter's body, the faster the tip of the bat will be moving when it meets the ball. A player six feet eight inches tall would have a swing arc about 15 percent longer than Derek Bell's for the same given bat size. This translates to a bat speed of 109 mph based on arc size alone.

Of course, that extra speed is not free: You have to add more power in order to maintain the same rpm with longer arms, a longer bat, a longer golf club, etc. Otherwise, we could make the bat twelve feet long and send the ball from the new Yankee Stadium clear into the old one across the street. This is obvious when you feel a significantly stronger in pull on your hand when you swing that rock on a two-foot string than you did when it was only a foot long.

So the second thing we need to do, in addition to giving our super hitter long arms, is to give him the power to increase the rotational speed of his upper extremities. He needs to have very strong shoulders, latissimus dorsi ("lats"), and biceps and triceps muscles so that he can swing around the arc with great force. However, with greater muscle mass often comes greater muscle *resistance*, or loss of flexibility, so our player has to stop somewhat short of looking like the Michelin Tire Man. An increase in muscle mass without counterproductive loss of flexibility would be about 20 percent over Derek Bell's playing weight of 215 lb, height of 74 inches, and estimated body fat of 9 percent.

2. FORWARD STRIDE

This occurs when the batter shifts his weight from the back leg to the front, much like a golfer. You can demonstrate this to yourself easily: Stand up and take a batter's stance as though you were expecting a pitch. Then start to swing the bat and put your body into it. You'll feel your weight transferring to your front leg, and you'll also feel your upper body moving forward. Getting your body mass behind the swing adds power but, more importantly, the more distance over which you move that mass, the more momentum you can develop.

A batter can only develop so much rotational speed, but if he moves his body forward at the same time at, say, 10 mph, that extra speed gets added to the sweet spot's speed relative to the ball.

One of the reasons cricket bowlers throw much faster than base-ball pitchers is that they're allowed to take a running start before they let go of the ball.

Here's how it works from a technical perspective:

The hips begin the process by pulling the body forward toward the contact zone. When leveraged properly by the stance or posture, the hips can pull the body very violently forward creating about 8–12 mph of forward body speed. The back foot pushes forward and rotates in sync in order to complete the forward momentum phase. There's controversy surrounding whether the back side is dragged forward and there isn't really a push off the back foot, but either way, all else being equal, when the body speed is increased, there is an increase in bat speed. According to research conducted by Perry Husband, a former professional player who now owns the country's premier baseball school and lectures worldwide on improving hitting technique, this increase in body speed produces about 6–10 mph in the speed of the ball coming off the face of the bat.

The longer the batter's stride length, the more room over which he can accelerate his forward motion. So we want our ideal hitter to be tall but, according to Dr. Aish, "Not too tall: There's a point at which all that extra size starts to work against you." There are two reasons for that: The first is that it takes time for nerves and muscles to react and execute the required motions. The larger the muscles, the more time it takes. This is related to why there is a fairly clear correlation in the animal kingdom between size and muscle speed. Compare the twitchy movements of a hummingbird or goldfish against the more ponderous ones of an elephant or a whale, and you'll get the picture.

The other reason is simple physics: The time and effort required to swing a 35-ounce bat up to 100 mph in a five-foot arc is a lot less than in an eight-foot arc. So for both of these reasons, the Jolly Green Giant might be able to whack one over the wall at Wrigley but he'd have to start his swing while the pitcher is still in his windup. "The optimal height for stride advantage in regulation baseball," according to Aish, "would be 7.5 percent taller than Bell, or eighty inches." How convenient: It's the same height that bought us a 15 percent bigger swing arc. According to Aish, it now brings us another 4–7 mph because of increased stride length.

"Four or five more miles per hour is reasonable," he calculated. "Seven is the very upper limit."

The upper limit is what we're looking for: Our bat speed is now up to 116 mph.

3. Pelvis rotation

The quadriceps, hamstrings, abdominal musculature, and intrinsic pelvic muscles conduct a coordinated motion to rotate the lower body, providing additional rotational forces to the bat. We can't get a very big advantage here because the lower part of the body is very compact and so there are no arc lengths to increase, but the lower body does contribute torque to the upper. Dr. Aish estimates that the right physique would add 4 percent more power, bringing us up to 120.6 mph.

4. Wrist release

While all of this powerful body rotation is going on to try to get the bat accelerating to as high a speed as possible, there's still one more opportunity to impart a last burst of power. It's pretty much the same technique as a pitcher or golfer uses, and that's to "snap the wrist" at the last instant before release or contact. The snap is a fast twisting motion that uses the muscles of the hands and arms to further increase the rotational speed of the bat.

If you watch a pitcher, golfer, or batter carefully, you'll notice that his wrists are bent way back when he's at the top of the swing or throw. This is called "cocking" the wrist, and it's analogous to cocking a gun. What they're doing is putting their wrists as far back as possible so that when they release the pitch or hit the ball, they can bend the wrist sharply in the other direction. You can visualize it as the same kind of motion a fly fisherman uses when he casts: His arms are perfectly still and he whips the rod forward using only his wrists.

Which, by the way, are surprisingly strong. All that power being generated by a batter gets transmitted by the wrists, and that last snap can account for as much as 20 percent of the ultimate bat speed. In technical terms, the wrist is sliding from maximum radial deviation to maximum ulnar deviation, for a total of about 75 degrees in a typical human wrist arc. With abnormal conditions such as Ehlers Danlos disease or Marfan's syndrome, that could almost double, but in a lot of sports even "normal" athletes have unusual radial/ulnar deviations that help them do things like take a golf club back beyond parallel or impart more spin to a bowling ball. We're going to assume that our ideal hitter has a maximum deviation of 115 degrees, based on the structure of the bones, which precludes anything higher. That gets 25 percent more power out of the wrist snap than Bell does. Twenty-five percent more of a 20 percent contribution to bat speed gets us an overall increase of 5 percent.

Final bat speed? Nearly 127 mph.

We now have everything we need to calculate the perfection point for the longest ball it's possible for a human batter to hit using an official baseball. It's not an easy calculation. It involves a series of differential equations developed by Robert Adair for his book, *The Physics of Baseball*, that take into account the effect of wind (which we'll assume to be zero), launch angle, etc. The beauty of Adair's model is that it's not a simulation of some idealized laboratory environment but is a highly accurate predictor of what would happen if a real batter hit a real ball in a real ballpark.

Bear in mind that the current official record for the longest ball ever hit is a 565-foot bomb by Mickey Mantle at Griffith Stadium in Washington on April 17, 1953.

Coors Field, Denver, Colorado
Date: May 18, 2432

Some four hundred years from now, a slugger we'll call Smith, about six eight and weighing around 247 pounds, will step to the plate at Coors Field in mile-high downtown Denver, taking a few last practice swings while fixing the pitcher with an inscrutable stare. It's a warm day in Denver, and dry, too, with record low humidity that has gone on for weeks.

The catcher for the opposing Yankees, a savvy veteran, looks out above center field and notices that the flags, which had been snapping in the stiff wind earlier, have settled down considerably. Small favors. He calls for time and heads for the mound.

The pitcher isn't surprised to see him trotting out. The batter is one of the most feared in the league, the best power hitter in the history of the game. The pitcher turns his head and notices a white moon hanging in the blue sky right behind him. Looks like it's at the perfect angle to give the batter an aiming point. At this altitude and in these conditions? Hell, he might even hit it. Which is why the catcher is coming out to the mound. And the manager, too, along with the second baseman and the shortstop.

It's the bottom of the ninth with two outs and the Yankees ahead 2–1. The strategy is obvious: Walk the monster, who's batting cleanup, and then go to work on the journeyman who's up next and close this game out. It makes sense, but it's a shame, because the Yankee pitcher has a two-hitter going and would like to finish it out by striking the slugger out or forcing an easy grounder for a play at first. But he gets it, and is willing. Winning is what counts.

He's already nodding when the catcher reaches him and says, into his glove so his lips can't be read, "You know what we gotta do here." The catcher has to shout, because the noise from the delirious and expectant crowd is overwhelming. As far as they're concerned, the Mighty Casey himself is at bat, only this time he isn't going to strike out.

The pitcher is kneading the ball, trying to make it look like a worthwhile endeavor and not some nervous habit but knowing he isn't fooling anybody. When he feels the second baseman put a hand on his shoulder, he says, "Yep. Let's get it done."

The manager arrives and tilts his head toward the Rockies dugout. "They're getting ready to pull Harris and put Domenico up after Smith." Harris is good at the bat but Domenico is a .315 hitter who usually warms the bench until a late inning because he's not much of a fielder. His primary usefulness is as a designated hitter during interleague games.

But even though this game is in a late inning, the catcher is still surprised. "Thought he was injured."

"His leg, yeah, but he can still hit," the manager replies, "and they'll put in a pinch runner if he gets to base."

"So what're you telling me?" the pitcher asks, pointing toward home plate. "We gonna pitch to this guy for real?"

"Yeah," the manager says, then holds out his hand, palm up. "But not you."

Four sets of eyes whip toward him, but before the protests can begin, he says, "We all know this guy can hit a ton, but he only hits .260, on account of he doesn't see the ball so good. So what we're gonna do is, we're—"

"Rubino," the pitcher says as he finally hands the ball over, naming a rookie flamethrower being groomed as a short reliever who has yet to appear in a major league game.

"No way is Smith not going to swing if it's even close to the zone," the manager says. "If Rubino can hit the corners with heat, this game is over." He knows the rookie throws too hard to be good for more than one inning, but all he needs from the kid is one out.

It seems impossible but the roar from the crowd grows even louder as a door in left field opens and the new pitcher lopes onto the field, trying unsuccessfully to hide a grin. It finally fades as he nears the mound and realizes that his joy is at the outgoing pitcher's expense.

"Awesome game, dude," he tells the pitcher. He doesn't see the old-school catcher's eyes roll at the phrase.

The manager hands him the ball. "I don't care if your whole arm flies offa your shoulder," he says, not letting go of the ball until he has the kid's full attention. "I wanna see smoke comin' off this ball."

The rookie grins at him. "You got it, skipper."

His warmup pitches are nothing special. He grunts and grimaces with effort but only throws at around 80 mph. He knows the hitter is watching him carefully, and there's no sense tipping him off to what he's capable of. You only get one debut in the Bigs.

As planned, his last three permitted warmup pitches are well outside the strike zone, and he looks worried. He isn't: They went precisely where he wanted them to. He could do that with slower pitches.

Smith, sensing an easy kill, steps to the plate for the second time. He grinds his feet into the dirt as he takes his stance and throws his hardest, most intimidating look at the pitcher, who quickly looks away.

The catcher puts his right hand between his legs but doesn't give a sign. There's no need to. The pitcher shakes him off anyway, then does it again, telegraphing indecision and dissent to the hitter, who laps it up.

The pitcher's windup has nothing of the awkwardness he displayed during his warmup. He straightens up, takes a breath, then rears back twice as far as he had been doing. As he uncoils forward, his body looks like fifteen kids in a game of Snap the Whip all rolled into one, a single long arc of pure power with his right hand at the very tail end of the chain. The entire piece of choreography is designed to use every muscle in his body to propel himself forward so fast that the trailing right hand is slingshotted toward home plate as it struggles to catch up with the rest of him. At that exact moment he lets loose the ball, giving it one final nudge with the tips of his fingers to get one last measure of the momentum that separates fastballs from *fastballs*.

By the time Smith realizes that a laser-straight pitch with nothing on it is heading for the dead center of the strike zone, he's already heard the sound of the ball slamming into the catcher's glove. Strike one.

He's not happy. Who the hell is this cocky kid to throw a strike on his first pitch to the greatest slugger in the game? The suddenly quiet crowd doesn't do much for his self-esteem, either, nor does seeing 101 mph on the display board in left field.

No way he does this to me again, Smith thinks.

The kid does it again. This time it's 104 mph, and the atmosphere in the park shifts noticeably after Smith watches it go by. The fans still want their man to put one into the rafters, but they also sense they might be witnessing history, and they're conflicted as to what they want to happen next.

By now the wind has died completely, the stillness causing an eerie unease in the crowd as well as in Smith. He has to make a decision. He's not going to get a hit by watching what this kid's pitches are going to do and then acting on it. He has to guess, and then commit himself to that guess, no matter what.

At this point, it's a sure thing the pitcher is going to throw one outside the zone, which makes sense, but he hasn't done anything that makes sense so far. Could it be another fastball down the middle? That'd certainly make headlines, getting Smith out on three strikes.

And that's why they—the catcher and pitcher—weren't going to do it. They'd think Smith was expecting it, and they can't take the chance. They have to use a different pitch. But which one? The Yankees hadn't gone over the pitching reports for Rubino because nobody thought he'd be in the game. So what was the most likely pitch if it wasn't going to be a fastball?

Curve. Had to be.

This time it was Smith who feigned nervousness, not swinging the bat in as wide an arc as he had been when he stepped back before the previous pitch. Tension was all over his body as he took his stance once again. He visualized the pitch he was expecting and exactly how he was going to hit it. By not worrying about seeing it through its entire flight, he could concentrate on just "spotting" it before it arrived to make sure he'd meet it properly.

The rookie bent himself backward into a pretzel as he wound up, and this time the grimace was real as he poured everything he owned into his legs, hips, and right arm. Halfway through the pitch, with his throwing hand still behind him, he knew he was into something sensational and the resulting jolt of adrenaline felt like it was going all the way to the ball. By the time he released it with a powerful flick of his wrist, he already knew what the pitch speed indicator in left field was going to say.

Smith knew it, too, but he didn't care. He could tell right away he'd guessed wrong so it didn't matter that the ball left the pitcher's hand at 111 mph. He knew exactly where it was going to end up, and when, and he also knew that the fastball was a perfect setup for the rocket he was fully capable of hitting. As far as he was concerned, the faster, the better.

No athlete uses more of his muscles at one time than a batter, and Smith thought he could feel each one of his coming into perfectly orchestrated play as he started the bat around, poured power down into it, saw with perfect clarity how the sweet spot was going to arrive at the exact right place in the cosmos when the ball showed up there as well, and kept pouring it on with total abandon through the entire arc of the swing because there was no way this was going to be anything other than utter perfection.

The ball was still moving at over 100 mph when the bat, moving at 127, finally connected. The sound was something no one in the ball park could quite recall ever having heard before. Even in the upper deck it sounded crisp and sharp, like a rifle shot or a dry log breaking cleanly in two.

The ball left the face of the bat at 194 mph and soared upward at a 35-degree angle. Backspin caused it to rise sharply at first, and it was still heading upward when it rose above roof level, only starting its descent when it had safely cleared the north wall.

A young boy outside the park listening to the game on the radio looked up in time to spot the ball against the clear sky, hanging in the air for what seemed like forever before it started down. He watched in awe as it accelerated toward him and he fought the urge to try to catch the incoming missile. At least on the fly: He backed up to where he thought it would come down after bouncing. When it finally landed, 9.3 seconds after Smith hit it, the young boy saw that he'd misjudged it and had to run backwards furiously after it hit a patch of dirt on the sidewalk and rose back into the air. Two bounces later he had it, just as a bunch of other kids and some adults began running in his direction.

He didn't think anyone would believe him, so he went back to the patch of dirt on the sidewalk and found the spot where the ball had first come down. It wasn't difficult to locate: The ball had created a comet in the dirt, an image so sharply etched you could tell within two degrees the direction the ball had come from.

One of the adults pointed to it and said, "That's where it hit! I saw it!"

"Me, too," a kid said.

The comet in the dirt was exactly 748 feet from home plate.

In the final chapter, we'll talk about the likelihood of our various perfection points actually being achieved, but we can pretty much dispense with this one right here. Whether baseball players will ever get to a sufficient level of physical prowess is beside the point, because there are three reasons why Major League Baseball is not likely to let things get to where anyone is going to be belting 700-foot homers into the streets around ballparks.

The first reason is that, much like the basketball slam dunk, the big league home run is already in danger of becoming just another in a growing list of once spectacular, now mundane feats. A lot of factors contributed to what has become a dizzying rise in home run production. The most obvious is that ball players, like all other athletes, get better as time goes by. The explosion in compensation, especially for super sluggers, has led more players to pursue rigorous fitness programs. Pitchers get better, too, but at a slower rate, and there are fewer ways for them to improve. There's also the fact that there have been some major advances in recruiting, including not only African-Americans but players from all over the world, particularly Latin American and Asian countries. With a much larger population to choose from, the talent level is going to be much higher.

In addition to those "natural" factors, we also have to take into account things that the ruling powers in the game have done to boost offense, including the elimination of trick pitches like the spitball, introduction of the designated hitter in the American League, lowering the pitcher's mound and shrinking the strike zone in 1969 and, until recently, ignoring the use of steroids. Subtler changes included the evolution of bats from hickory to ash to ash/resin composites to maple, along with ballparks swallowing up outfield space in favor of more paying seats.

The result? In 1907, there was one home run hit about every 300 at bats. A hundred years later, it was ten times as many. Here's a graphical look at the average number of home runs per game over the last 108 years:

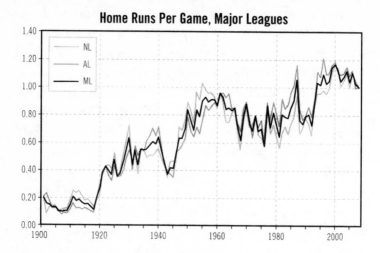

Home Runs Per Game, Major Leagues

Legend: NL, AL, ML

While many fans love to see a homer, too many of them dulls the excitement. At some point, the front office of the sport is going to have to tamp things down a little or risk a serious dilution in fan interest.

The second reason the sport will be motivated to keep long-ball hitting in check is out of fear of making ballparks obsolete. In the same way that the U.S. Golf Association struggles to keep long drives reasonable so as not to render existing golf courses too short, the game of baseball is going to need to make sure that owners don't have to surrender valuable revenue-producing seats in order to push the outfield boundaries back. There's already talk of doing just that in the new Yankee Stadium, owing to an unusually high number of balls sneaking over the wall. Dubbed the "ballpark on steroids," the replacement of the House That Ruth Built seems to contain an invisible wind tunnel. The first three games ever played there resulted in 17 homers. The first five games saw 25. In the entire 2008 season in the old stadium, there were only 160.

Owners are going to get pulled in two directions on this issue.

On the one hand, they believe that homers put butts in the seats, they don't want to have to give up any of those seats even to prevent a developing travesty, and they also aren't about to spend millions altering ballparks that took hundreds of millions to construct in the first place. On the other hand, they can't afford to sit idly by while the game becomes a parody of itself. The solution lies in fine-tuning other factors to make sure that there are only a reasonable number of homers hit without altering the stadiums. Whatever those adjustment factors are, they're going to prevent anybody from walloping the kind of long balls that might get them close to the perfection point.

Even if those two reasons for limiting long-ball hitting aren't persuasive enough, there's one more that is: ensuring the safety of pitchers.

A line drive leaving the face of the bat at 160 mph will reach the pitcher about a quarter of a second after being struck, which is not enough time for him to get out of the way. He wouldn't even be able to get a glove up to protect himself, because he won't see the ball coming in time. At that speed, a ball that hits him in the head could kill him instantly. If it struck him in the chest, it could stop his heart; it happened to a fourteen-year-old Little Leaguer in 2006, and the ball wasn't going anywhere near 160 mph. In 2000, Red Sox pitcher Bryce Florie took a line drive in the eye that fractured several bones in his face, requiring extensive surgery. (He came back a year later and pitched seven games but was released midseason.)

As hitters get stronger and bats keep improving, it's inevitable that a big league pitcher is going to get even more severely injured than Florie. As long as it's a rarity, there's not much likelihood that steps will be taken to prevent further occurrences. But let two or more such incidents take place within a relatively short time, and a call to arms will arise. Not only that, but pitchers will start to get skittish and tentative, which will make their pitches easier to hit, thus compounding the problem.

There's already evidence that the owners recognize the need to keep long balls in check. The most persuasive is that the major leagues have so far not adopted the aluminum bat that's used in college ball. While some might maintain that this is because it would overturn a century and a half of tradition, that's not a very credible argument: If the owners thought it would bring more fans to the park, they'd allow hitters to use grenade launchers.

The real reason is that the ball comes off the metal bat too fast. Big league hitters would turn a high percentage of shots that would have been simple fly outs off of wooden bats into fence-clearing home runs, destroying much of the elegance of the defensive game.

Right now, despite objections from some fans, there's a fairly good balance in the game between offense and defense. A top-notch pitcher is still capable of giving hitters a ton of trouble, but at the same time there is an average of 9.2 runs scored per game (up 7 percent from thirty years ago), which is enough to keep offense lovers happy. It took a lot of fine-tuning to get there, and while there are still too many homers being hit for some, it's a good bet that the keepers of the sport's flame aren't going to let it tip too far in either direction.

While that's good for the game, it does mean that we're not likely to ever witness anyone actual delivering the perfection-point home run.

CHAPTER NINE

THE MARATHON AND THE MILE

The limits of the legendary distances

They're the two best-known foot races in the world. One held the entire world captive over half a century ago. The other is Everyman's Everest, motivating and vexing us in equal measure. Both have provoked endless speculation about the ultimate in human performance. Can we apply some hard science to the argument and end those debates once and for all?

April 14, 2245
Gazi, Kenya

As he took what he knew would be a last look at the walls of his modest home, Zhafed Kenjobo marveled at how quickly life's fortunes could shift even as his ears picked up the sounds of distant but approaching gunfire.

Barely six weeks ago he'd stood atop the endurance sports zig-gurat, having defended his title at the Moscow Marathon and set yet another world record, only the latest in a string of extraordinary running achievements. Now, yet another sloppy and ill-considered coup in his native Kenya threatened to overturn his world.

He'd built this house exactly 26.2 miles from the Tanzanian border, an inside joke in the running community, as it was the same distance as the international standard marathon. Every Sunday morning he'd run to the border where Tanzanian customs agents, among his most ardent fans, would cheer his last steps and then drive him home, the two countries having had cordial, open-border relations for years.

Until the latest coup. Surprising in its swiftness and appalling in its brutality, this one had shocked the world into cutting off relations with the self-styled "freedom fighters." Isolated and desperate, the leaders of the coup retaliated by forbidding Kenyan citizens to leave the country. Those who hadn't taken the edict seriously had suffered terribly, and for their own protection, Somalia and Ethiopia to the north had closed their borders, as had Uganda in the west, preferring to deny humanitarian access rather than be party to the slaughter of refugees. Tanzania had been the last holdout, and it was to that border that Kenjobo had sent his wife and two small sons ahead of the rebel battalions that were advancing on Gazi. One man had no chance of fighting them off, but it was his hope that, being the most famous citizen of Kenya, he might stall them long enough to give his family enough time to make it to the border. That they'd let him go after that was out of the question: They'd rather kill him than have him out in the world telling the truth about their vicious tactics.

But the troops should have arrived already, and as the minutes passed, Kenjobo started to entertain the wild notion that he might make it out himself. All the vehicles in the village were gone, but he had his running shoes on and had spent the last fifteen minutes stretching and staring at his phone. The border guards would call him as soon as his wife and sons had made it safely across. Until then, he didn't dare leave. The gunfire sounds were growing louder, and he could now pick out the distinctive and terrifying crackle of "smart" bullets that were deadly accurate within 100 meters. Another few minutes and his life was as good as—

The electronic bleat of the phone startled him. As soon as he hit the button he heard a single word: "Go!"

"My wife and sons—" he started to say.

"They're safe! *Go!*"

Kenjobo went. As he stepped out of the house he realized that the insurgents were even closer than he'd thought. Out of force of habit he walked to a white line drawn in coral stones across the far side of his backyard. From that line to the border had been measured by the world governing body and determined to conform precisely to the official marathon distance. Kenjobo put one foot on it and, again out of force of habit, took three deep breaths, hit the START button on his watch, and began running.

He'd thought about this in the hour since his family had left for the border in one of the last electric trucks to leave the village. If he had the luxury of time, he'd run at his world record pace, but if the thugs nearing his village caught wind of his escape and came after him, he'd have to make adjustments.

The first three miles calmed him, as they usually did. It had always been his style never to look at what his competitors were doing in the early miles because it didn't matter. If they were well ahead of him, they'd gone out too fast and would fall apart later. If they were behind, maybe they'd catch up but there was no way to know and therefore nothing to be done about it. He was the world record holder and had broken the mark himself seven times, so it wasn't likely that he was going to be surprised. Only later, in the final few miles, was there the possibility that a heretofore unknown outlier might challenge him.

Early in the run like this, when he was in fully aerobic metabolism so that his glycogen stores were breaking down easily into energy-giving ATP, he felt as though he could fly. Less seasoned runners often let themselves get carried away with the near-giddy feeling, but Kenjobo was way past that kind of mistake. Later in the run, when his glycogen supply got low and began breaking down into lactic acid instead of ATP, he'd be thankful he'd pushed that painful threshold back.

At the seventeen-mile mark, the sun was setting, and as shadows began disappearing, he dared a look back over his shoulder. There in the gathering twilight he saw headlights shimmering atop the hill that formed the northern boundary of Gazi. It would be less than ten minutes before they reached his house, and only a few seconds more before the soldiers of the makeshift battalion figured out that he wasn't there.

The temptation to speed up was almost overwhelming, but Kenjobo knew it was also suicidal. It would do him no good to buy a few more minutes or seconds worth of distance now just to run out of gas before covering the crucial last miles to the crossing point. He forced his mind to stillness so he could concentrate on following the plan that would get him to the border in the shortest amount of time possible. He looked back once again and this time saw red taillights mixed in with the yellowish headlights. He smiled to himself, picturing the confusion that had all of those vehicles scattering to find him.

At mile twenty, another look told him that the confusion was over. At least three sets of headlights were aimed squarely down the road, pointing straight at him. It wasn't the best of roads, but the soldiers could still make 50 mph, maybe more. They could reach this point in less than twenty minutes. By then he'd be nearly five miles farther up the road but, even so, the numbers did not look good. Kenjobo assumed that the rebel troops wouldn't fire on him if he was in sight of the crossing—his Tanzanian friends wouldn't hesitate to vaporize them if they did—but he'd have to get within less than a mile to be safe.

Ten minutes later he heard a soft splat about fifty feet off to his right but he ignored it. A minute later there was another one, to his left this time but closer. Then another. Then something struck the asphalt ahead of him, making a bright spark in the dying light.

Bullets. The insurgents were lobbing them in his direction despite the distance. Were they hoping to get lucky, or to distract him?

It didn't matter. The point was that they'd spotted him. Just as that thought entered his mind he heard the sounds of their engines, faint but unmistakable, rising and falling in pitch as the vehicles bounced over potholes and ruts in the road.

Four miles left. By this time lactic acid was fully suffusing Kenjobo's muscles and even the foil gel packets he'd been sucking down weren't doing him any good. He was burning fat now, and it wasn't going to get any better.

His plan was no longer going to work. Pacing himself was out of the question. As the engine sounds grew louder in his ears he knew that, in order to succeed, he'd have to risk failing. There was no middle ground, no compromise.

A bullet struck the ground less than twenty feet behind him. Kenjobo, out of options, ramped it up a notch and within a minute was assaulted with fresh pains in his quads and a deeper ache in his lungs.

He was a running machine, ideally conditioned for both the efficient delivery and maximal utilization of oxygen. He was able to store vast quantities of carbohydrates relative to his weight, and his muscles knew how to burn fat when they needed to. Even so, there was no escaping the laws of physics: At some point there would simply be nothing left.

But he wasn't there yet. He'd felt this kind of pain before and knew he could take it and keep up the pace for another two miles. The only problem was that he had three miles to go to get himself within sight of the border guards.

A few minutes later, some of the bullets were landing within ten feet of him, and he thought he could make out shouts down the road behind him. At the twenty-five-mile mark, the vehicles were close enough that he could see reflections of their headlights playing in the leaves by the side of the road. But he could also see a glow coming from the border station, and knew that he would see it after one more left-hand bend in the road.

He sped up. Seconds later the sharp sounds of more conventional gunfire rang out, and rounds slapped into leaves and tree trunks. As he rounded the bend and caught a first glimpse of the border, the cessation in shooting he'd expected didn't come. If anything, it was increasing. He hadn't expected to slow down at this point, but he hadn't thought he'd need to speed up. He was in raw agony now and doubted he could hold on for much longer, but if he stopped, he died.

Suddenly the darkness gave way to a flood of light coming from the towers standing sentry at the border. Angry shouts from behind him mingled with encouraging cries from in front as the guards saw him. They heard the gunfire, too, and seconds later they began firing back. With rounds whizzing past his head from both directions, Kenjobo knew that it was only a matter of time before one of them found him. Both sides were firing blind, but as soon as the rebel vehicles rounded the bend and came within sight of the Tanzanians, it would become a full-scale firefight that wouldn't let up until he was officially off Kenyan soil.

Five hundred yards to go. A tracer from one of the towers streaked over his head, followed by a laser-guided RPG. A second later, a horrific explosion lifted a rebel vehicle off the roadbed and sent it crashing into the thick vegetation. A piece of shrapnel brushed past Kenjobo's arm. He saw people up ahead beckoning to him frantically as others fired from behind them to provide cover.

A hundred yards. Machine guns were being deployed now, the firing becoming wild and desperate. Kenjobo was too crazed and drained to even notice a Tanzanian sentry go down. His reserves were utterly depleted, his spent legs fueled only by the fear that at the next moment a molten round would enter his back.

As he came down to the last few yards, the Tanzanians, stationed in towers above and therefore no longer worried about hitting him accidentally, let loose a barrage of machine-gun fire, RPGs, and smart bullets. As arms reached out to grab Kenjobo, whatever autonomic part of his brain that was still working caused him to lift his right hand and stab at the STOP button of the watch on his left wrist.

Later, when the firefight was over and he was able to look at the time displayed on the watch, he assumed he must have made a mistake somewhere. But the border guards had a to-the-second record of the call they'd made to his house and surveillance equipment that recorded the exact moment at which Kenjobo had hit the border gate.

There was no mistake.

The Marathon

Exactly 2,500 years ago, in late summer of 490 B.C., a young man named Pheidippides did his first and last marathon. He probably wasn't in the best of shape, having just fought in the Battle of Marathon, so after running twenty-six miles from the battlefield to Athens to announce that the Persians had been defeated, he promptly collapsed and died. Whether this actually happened is up for debate, but following the 1879 publication of Robert Browning's poem commemorating the event, the story became firmly embedded in legend, if not in fact.

When the Olympic Games were resurrected in 1896 following a 1,503-year hiatus, the organizers scouted around for a special event that would popularize the Games and recall the glory of ancient Greece. They decided to restage Pheidippides' epic run, beginning with a qualifying race some months before the Games themselves. The winner of that race, the first marathon of modern times, was Charilaos Vasilakos in a time of 3:18. That's hours and minutes. Nobody thought there'd be much point in taking note of seconds. By the time of the main race during the Games them-selves, the world record was reduced to 2:58:50, and seconds were introduced.

Except . . . the route was only about 40 kilometers, or 25 miles, instead of the modern 26.22 miles, so a comparable full-length record would have been about 3:07. The distances weren't fixed in those days, the thinking being that, as long as everybody ran the same course, what difference did it really make?

They tried to keep it close to the distance from Marathon to Athens, but the exact length depended on the specific route estab-lished for each venue. Many attempts were made to standardize the distance over the next dozen years, but most of those efforts focused on nailing 26 miles as the official length from the start line to the stadium. What happened once the runners were inside the stadium was another story. In 1908 it was decided that 385 yards around the track at the finish would nicely accommodate spectators, including Queen Alexandra, and that seemed to be the end of it. Then it was changed again in 1912, again in 1920, and four years later it was changed back to the 1908 distance. So the first seven Olympic marathons were run at six different distances. But it's been fixed at 26 miles 385 yards for over a hundred years now and isn't likely to change, because big money has entered into it, and when you can win a brand-new Mercedes for breaking a record, inches matter.

The marathon has progressed from being the exclusive pur-view of a handful of endurance weirdos to one of the most popular amateur race distances in the world. While the Olympic marathon is still limited to a small field of official entrants, nearly all of the more than five hundred other marathons run each year comprise large numbers of amateurs in addition to the professionals. These events have become known as much for their massive fields as for the pro competitions. The New York City Marathon has so many runners—forty-four thousand in 2009, same as the population of metropolitan Olympia, Washington—that two separate start areas are needed. By the time the leaders are four miles into the race, there are still competitors who haven't even reached the *start* line.

We've come a long way since Charilaos Vasilakos's first world record. His (equivalent) time of 3:07 would have put him in sixth place at the last New York City Marathon . . . in the men's 60–64 age group. The winner of the women's 50–54 division would have walloped Vasilakos by eighteen minutes.

The current world record is a mind-boggling 2:03:59 run by Haile Gebrselassie (it's pronounced "GEBra sell-AH-see," rhyming with "Debra's a lassie"), the legendary Ethiopian who also ran the second fastest time ever, 2:04:26. This comes out to roughly 4:43 per mile. That's only fifteen seconds slower than the first world record for a *single* mile in 1852.

In trying to determine the fastest marathon possible, it helps to understand a little of the science involved in pushing a human body to run twenty-six nonstop miles. The first consideration is where the energy comes from. Despite all the flowery (and true) talk about motivation and reaching deep and racing with heart, there's simply no getting around the laws of thermodynamics. To propel a body forward takes energy, and once the energy is gone, the running stops.

The energy required to move your muscles and thereby do some useful work comes from a substance called adenosine triphosphate, or ATP, which is generated by the breakdown of food. In endurance runners, the main energy sources are carbohydrates, stored in the body as glycogen, and fat. When there is plenty of oxygen available for metabolism, glycogen is easily broken down in a process called aerobic metabolism to create ATP which keeps you happy and moving.

But there's only so much glycogen you can store up, and when you start running low, you start relying on fat. Fat takes more oxygen to produce energy than glycogen does, and even then, it will only do it if there's still some glycogen left. If the glycogen gets too low and there's also not enough oxygen available, you start producing lactic acid instead of ATP.

Lactic acid is not good. It hurts. Beyond a certain threshold fatigue sets in and, even worse, glycogen starts breaking down even more quickly into lactic acid, creating a self-perpetuating cycle of progressive awfulness. Among marathoners, this is known as "hitting the wall," and it typically happens around the twenty-mile mark. It's possible to get past the wall, but only if you somehow kick-start the metabolism of fat. One popular way to do that is by loading up on easily absorbed carbohydrates from little foil gel packs. For many people, knocking back a few cups of chicken broth accomplishes the same thing. Gel packs and broth are commonly handed out at aid stations along marathon routes.

This little science lesson gives us some clues in trying to figure out what our ideal marathoner would look like. We can break it down into three specific requirements.

The first is optimal aerobic metabolism to break that glycogen down into ATP instead of lactic acid. While there is certainly a genetic component to that ability, most of it comes from training. The more you train, the more you increase your ability to both consume and effectively utilize oxygen. You get more and bigger mitochondria, which are little energy factories in your cells, along with more of the enzymes that assist in aerobic metabolism. Your heart and entire respiratory system get better at both pumping oxygenated blood to your muscles and making more oxygen available when it gets there.

The second attribute of the ideal marathoner is that his muscles are highly efficient at utilizing fat, which goes a long way in preserving the amount of carbohydrates he retains. Training is a key factor here as well, because the same techniques that increase efficiency in the use of oxygen also help muscles to deploy fat.

The third ideal condition is the ability to store up more carbohydrates to begin with, except not in your belly in the form of just-eaten energy bars but in your muscles in the form of glycogen. A common and somewhat discredited technique for doing that is "carbo loading." The idea is to cut down drastically on carbohydrate intake a few days before a race, reducing the amount of glycogen stored in the body. When you get to the point where the body "cries out" desperately for replenishment two days before the race (if you timed it just right), you start ingesting huge amounts of carbs. The theory is that the starving muscles will end up soaking up a lot more glycogen than they normally would, as though they were afraid of running low again. This probably attributes a lot more intelligence to muscle fibers than they actually have, but "peaking" in this manner still has its die-hard adherents.

Again, some of those ideal attributes have a genetic basis, but some can be developed via training. Most world-class professional marathoners have training regimens that many of us would view as extreme—or bizarre: One world-class pro actually did all of his training on a quarter-mile track—so it's not likely that anyone is going to approach the perfection point for the marathon by out-training everybody else. But some kinds of training are better than others, especially if the athlete is genetically preconditioned to take maximum advantage of it.

One thing that we're not likely to see in someone pursuing the perfection point is a great marathoner knocking himself out over a long period of time to get there. Surprisingly, while distance runners are obviously specialists in endurance, longevity seems to be a weak point. A 1985 study examined career performances of 25 athletes who met at least two of these three criteria: won at least one of the world's premier marathons; was a sub-2:09:30 runner;

and had been named as a favorite in the 1984 Olympic marathon. The results were unexpected:

- 70 percent of the athletes had their first victory in a significant world event within their first three races. All but one achieved victory in their first five races.
- 83 percent had their best times within their first five races.
- The majority of the athletes competed in only 10–12 quality races.

Although this has changed since the original study because of the money motivating professional marathoners to run more races, the researchers drew several conclusions that are still valid. One was that there seemed to be fairly severe limits on the number of highly competitive races an athlete can undertake. Something important seems to break down with each all-out effort, and it's not just physical but mental. The most notable effects of psychic wear and tear usually come in the last three to six miles, when the marathoner just can't seem to rouse himself to the levels necessary to win.

All of this is by way of saying that, while there's nothing to prevent marathon times from getting faster overall, there's a lot that makes it difficult for individual competitors to patiently improve over time. Some of it is psychological: It's just too grueling an avocation within which to stay fresh after more than a handful of competitions.

That brings us to another reason why we're not seeing great improvements in times, which is also true in other exceptionally taxing endurance sports, and that has to do with commercialism, at least on the elite international stage. Most of a professional athlete's earnings don't come from prize money; they get paid by corporations that make clothing, soft drinks, running shoes, nutritional supplements, computers, cars, and razor blades, or that provide services like accounting, travel, and banking. Sponsored athletes are far more interested in winning and therefore making more money from both prize purses and endorsements than in breaking records, because to break a record you have to risk losing, and money-oriented athletes are increasingly less willing to do that when sponsors are banking on them for victories.

This is especially true for athletes from poorer countries, some of whom are literally running for their lives. When Third World participants stare forlornly into the camera and talk about using prize money to buy some cows for the village, we tend to laugh, but it isn't funny. They're talking about their survival. You don't find a lot of Kenyans or Ethiopians in the Ironman or the Tour de France because those sports require that some serious money be invested in equipment and travel before any of that investment starts paying dividends, if it ever does at all. But running is free, and that's one of the reasons so many great track stars come from humble circumstances. For these people, as for the sponsored elites, winning is everything, and unless some competitor is right on his tail pushing him, a marathoner whose livelihood is dependent on victory and who knows that he's only got a very limited number of events in which to make his mark isn't going to risk burning himself out in the last few miles to try to break a record.

Money weighs against records in another way as well. Sponsors count on their elite athletes to be as visible as possible. They want them out there racing, which puts pressure on them to do as many races as possible. So endurance athletes try to maintain a fine line between competing often but not beating themselves up too badly. This results in slower times across the field, and it's also why there's a new trend in long-distance events whereby a top athlete who is behind in the race and realizes that he isn't going to win simply drops out, saving his body for the next time when he'll have a better chance. This has become fairly common in the world's toughest endurance event, the Ironman.

Our job is to ignore the foregoing, set aside all the commercial realities, and continue to the ultimate marathon performance. But it's not that easy, because endurance events have a mental component that is inseparable from the physical. You might think that physics is physics, and if the body is out of fuel, that's the end of it because there's no getting around the laws of thermodynamics. You'd be right, except for one thing: The body *never* runs out of fuel until it's dead. A marathoner who poured everything he had into his race and collapses across the line might look like he couldn't possibly move another muscle, but he can do plenty more than that. And if the line had been another quarter mile away, he would have gone another quarter mile and then collapsed. Ultra marathoners routinely run 100 miles at a stretch, and one of them, Dean Karnazes, once entered a 200-mile relay race and ran it solo against teams of twelve.

Give that collapsed marathoner five minutes, even without any water or food, and he'll be back on his feet and on his way to the press conference, using only the fuel he had when he crossed the line. Amazingly, an elite marathoner who just completed a world record race and collapses over the line only used about 22 percent of the available energy in his body. Where's the rest?

In his head, at least metaphorically. If a lion were to appear out of the bushes in Central Park, that half-dead marathoner would be off like a shot and wouldn't stop until he was safely ensconced in the Plaza Hotel on Fifty-ninth Street.

Even if he's in too much pain to run, "too much" is purely a function of his mind-set. The great runner Steve Prefontaine once explained his success by saying, "I can take more pain than anybody."

So in addition to pondering the ideal physiology and race conditions, we're going to have to make some assumptions about the mental strength of our ideal marathoner and the amount of available power it's reasonable to assume he can summon up in running the perfection-point marathon.

We can start our search with a question that has been intriguing people in the sport for years: Will anyone ever run a marathon in under two hours?

The "Sub 2" would seem to be as much of a Holy Grail to marathoners as the four-minute mile was to track runners, but it isn't: Nobody is currently trying to achieve it, because it's too far out of the realm of current possibility. When Gebrselassie set his first world record in 2007, he predicted a 2:02 "sometime in the near future." The following year, when he set the current record and broke the 2:04 barrier by one second, he wasn't quite so optimistic, and said instead that a 2:03:30 was possible. The prevailing sentiment within the marathon community is that a Sub 2 may be possible but is not likely to happen in any of our lifetimes. It's hard to know whether anyone actually believes that or if they just don't want to be remembered as short-sighted, in the manner of those who doubted that anyone would ever climb Mt. Everest or walk on the moon or get at least one hit in 56 consecutive major league baseball games. (That last is Joe DiMaggio's record, and it's number one on just about everybody's list of existing sports records that will never be broken.)

The best traditional predictor of marathon records isn't a physiological factor like oxygen uptake capacity or lactic acid tolerance. It's performance in 10K and 5K races. Trends in world records in those events tend to signal similar patterns in the marathon. This has to do with the fact that these races are tightly linked in terms of who competes in them. Gebrselassie and his fellow competitors took 45 seconds off the 10K world record in the mid-nineties and

18 seconds off the 5K. During an overlapping five-year period, the marathon record was reduced by a correspondingly large 1:10. Since then, however, records across the board have pretty much stabilized.

Because of the close correlation between 10K and marathon times (more about this later), for someone to run a Sub 2 marathon he'd have to be capable of running a 10K in somewhat less than 25:30. The current record? 26:17.53, and it's been there for over four years. There are another 48 seconds to be shaved off, and in the prior ten years, it came down by only 24. Needless to say, though, the effort required to set a new record is nonlinear as the record gets shortened: Knocking off the next 12 seconds is going to be exponentially harder than it was to lop off the last 24. How hard? Kenenisa Bekele, arguably the greatest all-around distance runner ever and the current 10K world record holder, dropped that record five seconds in ten years.

Another good place to look for predictors of marathon performance is the half-marathon distance. Elite athletes can generally run a marathon in about 6–7 minutes more than twice their half-marathon times. Therefore, in order for us to see a Sub 2 marathon, we're going to need to see a 56-minute half marathon first. Right now the record is at 58:33, and the gulf seems unbreakable.

Again, commercial interests play a role. In order to break a record, conditions have to be dead perfect: a course that hugs the edge of legality in terms of downhill component, ideal temperatures, and no wind. Few if any of the big-paying races look like that, so for an athlete to attempt a record run, he'd have to do a lesser event and thereby compromise his chance of doing well in a more lucrative race, because you can't do a lot of these in a year. (Remember what we said earlier about marathoners hitting their best races within the first three to five attempts?)

And the odds of getting a record in a smaller event are pretty slim, because without a handful of big-name runners pushing the leader to an extreme performance, achieving a top time is very difficult.

At this point, we have to couch our quest in the form of two questions. The first is whether someone is likely to ever actually run a perfection-point marathon, given the practical realities of the sport.

The second, and the only one of the two we care about for purposes of this book, is what the time of that marathon would be, assuming ideal course conditions and a race environment that would play directly into the mental requirements of the ultimate performance. It would have to be a winner-take-all race with a multimillion-dollar bonus for setting a record, with a field consisting of the world's best marathoners whose personal record times are all within a few seconds of the others', on a dead straight, point-to-point course having the maximum elevation drop permitted by international standards.

So what would our ideal athlete look like, and what's the fastest he could ever run?

The marathon is far too complex an undertaking to lend itself to easy analysis. There are many variables involved, and optimizing some of them runs the risk of compromising others. Our research demonstrates conclusively that there is no single ideal combination; there are lots of ways to get to the same result. However we do that, though, we know for sure that our athlete will need to possess these characteristics:

He will be as light as possible without being malnourished or unhealthy. The less weight he has to carry, the less oxygen he'll need to maintain a specific pace. In addition to overall low weight, we'd want him to have legs that are light in relation to the rest of his body, using the thoroughbred horse as the model.

It's interesting that lung capacity is not a major factor: There is no correlation between how much air the lungs can hold and how fast someone can run. Faster breathing takes care of any shortages of air, and you only actually use a fraction of the available oxygen in the lungs anyway; more of it is exhaled per lungful than is actually absorbed. The issue with oxygen isn't how much you can get into your lungs but how efficiently you use what's in there. Delivering it to the muscles is what counts, along with how well the muscles make use of it when it arrives. Transport is enhanced by a lot of hemoglobin and a high percentage of red blood cells, which can be built up over time by the right kind of exercise. (As with our static-apnea athlete, living at high altitudes helps.) In order for the muscles to be able to use as much of that oxygen as possible, we're looking for high levels of mitochondrial density and aerobic enzyme activity.

Other ideal characteristics include a large glycogen storage capability, the lowest body-fat percentage possible along with the highest effective use of fat in order to delay glycogen depletion, and a metabolism optimized for making new glucose from noncarbohydrate sources via gluconeogenesis.

Once all of those elements are in place and in optimal balance, we can tackle the question of the fastest marathon possible. There are three different ways to come at that problem.

The first is a statistical approach that will tell us the fastest marathon that current champion runners are theoretically capable of. Examining what it would take to go under two hours is a good place to start. A lot of people who are knowledgeable about the science behind long-distance running believe that we're very close to the perfection point right now, and the distinct slowdown in improvements over recent years would seem to bear that out. At the very least, they believe that further improvements will be much less dramatic than they have been, and if even a 2:02 were to be run sometime in the next twenty years, it would send shock waves through the sports world.

As we said earlier, there are some very reliable rules of thumb telling us that, for someone to run a marathon in under two hours, he would have to be capable of running about 56 minutes for the half-marathon. The current record of 58:33 was set by Kenyan Samuel Kamau Wanjiru in 2007.

To understand the difficulty of running a half-marathon that fast, let's look at it in terms of how fast each of its thirteen individual miles would have to be run. Haile Gebrselassie's marathon world record of 2:03:59 equates to an average pace per mile of 4:43. This is about 51 seconds slower than what he can run for one mile if he's only running one mile. For him to go Sub 2 in the marathon, he'd need an average pace of just under 4:35, which is 45 seconds slower than his best mile. Nobody has ever come close to being able to do that. Nine seconds doesn't sound like much but lopping that much time off your mile pace for an entire marathon when you're already the fastest in the world would be like hitting a golf ball over an office building when the best you've ever done before is reach the clown's nose. Gebrselassie will never do it; no current world-class runner will.

What it will take is a world record holder in the mile with the genetic ability to adapt to endurance training and the willingness to make the switch to the longer distance. Let's see how that might work by using the current world champion in the mile, Hicham El Guerrouj (pronounced "ga-ROOJ," to rhyme with "la rouge") of Morocco. Gebrselassie's mile pace in his world record marathon was 23 percent slower than his best single mile. El Guerrouj's world record mile was 3:43. If he were to chuck his career as a miler and move over to the marathon, and if he could run that distance at 23 percent slower per mile than his best mile, he could conceivably finish in 1:59:52.

There are an awful lot of assumptions in that simple analysis, and as a practical matter, it could never happen like that. But if someone with El Guerrouj's talents and genetics were to undertake the marathon early enough in his career, and if that career was aimed at a single race of the kind we described earlier, a Sub 2 is conceivable.

To even postulate a marathon time of 1:59:44 would be considered insane by today's standards. That might be true—there's plenty of anecdotal evidence to support the argument that such a time is inconceivable—but even so, it's not good enough for us if what we're after is absolute perfection. The reason is that the analysis was based on the hypothetical performance of existing runners. El Guerrouj and Gebrselassie are real athletes, and the world records we cited are their real achievements. What kind of a marathon might we see from the best distance runner that could possibly exist?

Work done by François Péronnet and Guy Thibault at the University of Montreal in Quebec, Canada, provides an alternative approach to the physiology of endurance running that uses a set of factors based on metabolic energy-yielding processes. They looked at both the capacity of anaerobic metabolism and maximal aerobic power, but their key contribution was insight into how peak aerobic power gets reduced as an endurance event wears on. They developed equations that predict average power output for a large number of athletic events. When they tested them against actual performances, the average difference between those performances and their model was less than 1 percent.

Applying their model to the evolution of their factors along with the historical progressions of world records, they predicted that the first Sub 2 marathon will be run no later than 2028. By 2040, the record will be down to 1:57:18, and at some point before the human race evolves into a different species, it will be possible for someone to clock a dizzying but scientifically plausible 1:48:25.

I disagree. Their equations assume a degree of linearity that couldn't be sustained once marathon times dip much below two hours. A careful look at their results discloses that what they've really done is place what physicists call an "upper bound" on the marathon. Theirs is a limit that cannot be exceeded, but they haven't shown what could actually be done.

To explain the distinction with an overly simplified example, I think back to when I was a kid watching my father drive his car. The farther down he pressed the gas pedal, the faster the car went. I assumed that the car could go up to the highest number on the speedometer, because what was to stop it? Just keep stomping on the accelerator.

Of course, that wasn't true. The speedometer placed an upper theoretical limit on what the car could do, but it told me very little about what it could *actually* do. It's the same way with Péronnet and Thibault's numbers: Their equations provide an absolute limit on power output based on fundamental physics and physiology, but they don't say much about what a human can do in a real marathon.

For that we turn to Michael J. Joyner of the Mayo Clinic in Rochester, Minnesota. In 1991 Joyner examined three physiological factors considered to be the key limiters in human endurance performance. Using maximum, previously substantiated values for each one, he then correlated them with marathon running times in various combinations in order to find the mix that would result in the fastest time and still be within the realm of practical possibility.

The first factor was VO_2 max, a measurement of how many milliliters of oxygen are utilized per kilogram of body weight each minute. Generally considered the best indicator of cardiorespiratory endurance and aerobic fitness, VO_2 max (spoken aloud as "vee oh-two max") is the point at which oxygen consumption plateaus and therefore defines an athlete's maximal aerobic capacity.

The second factor is blood-lactate threshold, a point where lactic acid floods muscle cells too fast for the body to metabolize the excess. If you've ever felt "the burn," you've hit this threshold, beyond which speed, power, and efficiency start to suffer under the pain of anaerobic stress. Blood-lactate threshold is expressed as a percentage of VO_2 max, so if your threshold is 70 percent, you'll start to hit the wall when you're at 70 percent of your maximal oxygen utilization.

The third factor is running economy. Simply stated, this is about getting the most bang for the buck, i.e., the most forward movement for the least amount of energy expended. This one factor alone has a host of critical subfactors, ranging all the way from mitochondrial density to the shock-absorbing capability of the runner's feet to the way he swings his arms and the shape of his body. Perfect form requires as little braking action as possible: The foot hits the ground already moving backward relative to the ground so that there is as little deceleration as possible and a smooth transition from landing to push-off.

These three key factors interrelate in complex ways, but the basic idea is to balance them so that maximum speed is achieved at the blood-lactate threshold. Joyner determined that with exceptional but practically achievable running economy, the ideal combination is a VO_2 max of 84 milliliters of oxygen per kilogram of body weight per minute and a lactate threshold of 85 percent of VO_2 max.

According to his model, an athlete who achieves this ideal combination would be capable of running a marathon in 1:57:58. It's a far cry from Péronnet and Thibault's upper bound of 1:48:25, but it's nearly two minutes faster than the 1:59:44 we got from our simple statistical analysis.

1:57:58 is the perfection point for the marathon.

The Mile

Despite the mile's progressive obsolescence in an increasingly metric world, the distance has a hold on us that's difficult to shake. It's a cherished standard held dear for traditional reasons, celebrated in countless American songs ("Five Hundred Miles"), poems (". . . and miles to go before I sleep"), and popular expressions ("Go the extra mile"), far too deeply embedded in our culture for us to give it up, making the United States, the UK, Liberia, and Myanmar the only diehards still clinging to that cumbersome, well, *mile*stone.

Following World War II and our forced involvement with how the rest of the world measures distance, we had the opportunity to bring ourselves into line with a system far more wieldy, useful, and sensible. Those hopes were dashed, however, when an obscure young physician in England, after finishing fourth in the 1500-meter event at the Helsinki Olympics, set his sights on achieving a goal long sought by track athletes: running a mile in less than four minutes.

The idea that experts considered a sub-four mile impossible was largely a myth propagated by sportswriters eager to promote Roger Bannister's quixotic adventure. The world record at the time was Gunder Hägg's 4:01.3. Hägg and another Swedish runner, Arne Andersson, had run a series of head-to-head competitions in the early 1940s and succeeded in lopping five seconds off the record.

But there it stalled. While some attributed this extended failure in bettering the record to the disruption caused by the war, others point out that the two Swedes had broken it six times between them during the height of the conflict, from 1942 to 1945. Surely that was at least as disruptive a time as the postwar years.

The truth is, racing on the Continent was severely curtailed in the years following the war. Much of Europe lay in ruins and starvation was rampant. Although the Olympic Games were staged in London in 1948, no new facilities were built for them and the four thousand competitors who participated were asked to bring their own food. But 1952 saw a grander Games and, even though the mile wasn't run in the Olympics, it was nevertheless one of the most popular track events in the world, so the notion that sports was somehow still "on hold" after Helsinki was starting to wear thin as a rationale for the stagnant world record in the mile. By 1954, Gunder Hägg's nine-year-old mark of 4:01.3 still stood as the fastest mile. That's when people started to wonder if four minutes might represent some impenetrable barrier, just as they had once thought the sound barrier was impenetrable in flight.

There was no scientific basis for such a belief. While the sound barrier was a very real flight threshold with a significant set of physical obstacles requiring radical new designs to overcome, "four minutes" had no such built-in hindrances and wasn't of any more scientific note than 3:59.5 or 4:00.7. Or 4:01, for that matter, which also stood unbreached.

And yet, there was something tantalizing about being that close to a round number like four minutes. It made the quest for a new record more than just another in an endless series of obscure athletic achievements. It gave the public something identifiable to latch on to, and sportswriters the world over exploited the "barrier" in service of drawing attention to the track.

It couldn't have come at a more opportune time: A good chunk of the planet's population had just about finished crawling out of a painful postwar reconstruction only to find itself plunged into a Cold War with apocalyptic undertones that were difficult for the mind to grasp. In the fall of 1952, the United States detonated the world's first hydrogen bomb, completely obliterating an atoll in the Pacific with a force five hundred times that of the atomic bomb that destroyed Hiroshima. The next year saw the execution of convicted "atomic spies" Julius and Ethel Rosenberg, followed two months later by the detonation of the Soviet Union's first H-bomb. Duck-and-cover drills had become as common in schools as recess, and the world couldn't have been more ready for an uplifting, life-affirming, nondestructive and benign event than the fall of an athletic barrier that had been hyped up to mythic proportions.

And then, on a raw, windy day at the Iffley Road Track at Oxford, it happened: Roger Bannister clocked the mile at 3:59.4 and made front-page headlines the world over. He became the inaugural recipient of *Sports Illustrated*'s Sportsman of the Year award, and more than half a century later Bannister's is still a household name. No other individual athletic achievement comes close to his in terms of enduring public awareness. (By the way: The stadium announcer who informed the crowd at Iffley of Bannister's time was Norris McWhirter, who went on to create the *Guinness Book of Records*, so named because it was commissioned by the managing director of the famed brewery as a way of settling pub debates.)

As astonishing as this feat was, just forty-six days later Bannister's Australian rival John Landy brought the record down to 3:57.9 (officially recorded as 3:58 due to the rounding rules then in effect). In fact, the four-minute barrier was broken seven times that year. It took thousands of years for humans to evolve to a point where one of them could run a mile in under four minutes, but once the barrier was shattered, the seemingly unbreakable mark was broken as if it had never existed. Even so, bettering the record thereafter took another three years, and the improvement was only 0.7 seconds.

By 1967, American Jim Ryun had brought it down to 3:51.1, and there it stood for eight years. When that mark was cracked, it was by a minuscule 0.1 seconds, and it wasn't until the epic rivalry of Englishmen Sebastian Coe and Steve Ovett that records began to fall at shorter intervals. However, some of that had to do with the fact that the international governing body had begun measuring in hundredths of a second rather than tenths, as had been the rule until 1981. Coe and Ovett set new records five times between 1979 and 1981, but the total improvement during that period was only 1.6 seconds. That's an average of about a third of a second each.

In 1999, Hicham El Guerrouj set a new world record of 3:43.13 at the Golden League meet in Rome. That was eleven years ago, and his mark has not been bettered.

Track and field aficionados are beginning to wonder if it ever will be. Is it possible that the perfection point for the mile has already been achieved?

While it's true that there aren't very many mile-long races left in the world, most having been replaced by the 1500 meters, the fact is that the two are not that far apart. The mile is 359 feet longer, and virtually all athletes who compete at the mile also race the 1500. Because the two distances are so close, it's easy to equate one with the other. Multiplying an athlete's 1500-meter time by 1.0802 will provide a very close approximation of what his equivalent mile time would have been. (Based on pure arithmetic, you'd think the conversion factor would be 1.072, because the mile is 1.072 times the distance of the 1500. But it doesn't work out quite like that, because the athlete can't maintain his 1500-meter speed for the additional 359 feet needed to complete the mile. The 1.0802 factor was derived using curvilinear regression based on actual performances by athletes who have run both distances.)

As if further proof were needed that these two events might as well be the same, consider that the world record holder in the 1500-meter is Hicham El Guerrouj, the same man who also holds the mile record. He set them just a year apart, which means that the 1500-meter record has held up for twelve years, and it was on the same track in Rome where he would set the mile record. Using our conversion factor, his time of 3:26.00 equates to about a 3:42.5 mile, which is very close to his actual record of 3:43.13.

Bottom line, the argument that the mile record has stalled because of the dearth of competition at that distance simply doesn't hold water, because the nearly equivalent 1500-meter shows the same pattern. In the thirteen years following Steve Cram's 3:29.67 in 1985, the record fell by a grand total of just 3.67 seconds, to El Guerrouj's 3:26.00 in 1998, and there it has languished ever since. Why should that be, when world records in every other distance get broken with almost monotonous regularity?

The answer might lie in the unusual nature of the distance itself. While the mile is like an old friend to motorists and weekend joggers, it's hellishly alien if what you want to do is run just one. The mile is far too long for an all-out sprint and much too short to be considered an endurance event. Where 400-meter and 800-meter runners routinely push themselves to the anaerobic threshold, milers have to carefully monitor their oxygen uptake to make sure they've got enough left for the last few hundred yards. But where 5,000-meter and 10,000-meter runners are able to settle into a groove where they parcel out energy in closely calibrated doses, milers spend very little time in a steady-state zone. The race is far too tactical to allow for the kind of relaxed running that endurance racers strive for.

The "neither here nor there" nature of the mile has physiological implications as well. Most milers have plenty of raw energy stores left at the end of the race but are on the verge of fainting from lack of oxygen. That's why you see so many of them fall onto the track, heaving and gasping, as soon as they cross the finish line, whereas marathoners rarely do that despite spending more than thirty times longer out on the course than milers do.

In other words, the mile sits right on the cusp of two very different kinds of running, and therefore two very different types of training and physiology, and trying to combine the best of both worlds in a single athlete is extremely difficult. Too much of one type compromises the other, which explains why milers are such specialists. While many track and field athletes compete in more than one event—a good example of that is Usain Bolt, who holds world records in both the 100-meter and 200-meter sprints, two very different races—milers tend to stick to the mile and the 1500 meters. Training for anything else pulls them out of their mile-specific conditioning.

Can El Guerrouj's 1998 and 1999 records be broken? And if they can, what is the fastest mile it's humanly possible to run?

Physiologically, the mile is very different from the marathon. Whereas glycogen depletion is a large issue for the marathon and is one of the key limiting factors of performance, the mile is too short a race for competitors to run out of muscle glycogen.

The miler will also feel more winded at the end of a race and will be breathing more heavily than a marathoner. As we describe in the chapter on static apnea, the main stimulus to breathe is not an insufficiency of oxygen but too much carbon dioxide (CO_2), and there is a lot more CO_2 produced via metabolism when racing the mile compared to the marathon.

Fatigue in the mile is primarily caused by a high rate of an-aerobic metabolism, which occurs when you run too fast for your heart to provide sufficient oxygen to your muscles. Running faster comes down to increasing the rate at which ATP is resynthesized so it can be broken down to liberate energy for muscle contraction.

When you exceed your aerobic metabolic capacity to resynthe-size ATP, a number of problems begin to arise inside your muscles. They lose their ability to contract effectively because of an increase in hydrogen ions, which causes the muscle pH to decrease, a condition called acidosis. Acidosis has a number of side effects: (1) it inhibits the production of ATP; (2) it inhibits the enzyme that breaks down ATP inside muscles, which decreases muscle contractile force; and (3) it inhibits the release of calcium, the trigger for muscle contraction, from its storage site in muscles.

In addition to hydrogen ion accumulation, other metabolites accumulate when running fast, each of which causes a specific problem inside muscles, from inhibition of enzymes involved in muscle contraction to interference with electrical charges, ultimately leading to a decrease in force production and running speed.

While the effects of anaerobic metabolism cause that heavy, dead-legged feeling when racing the mile, limitations in aerobic metabolism also cause fatigue, by limiting the pace that you can maintain aerobically. Your legs feel like lead during the latter stages of the mile because you're not getting enough oxygen to them. That's why it's so important to put in a lot of miles during training even for athletes who run only one mile in competition: They have to develop themselves aerobically to delay acidosis and the accumulation of metabolites.

Fortunately, there is a way to cut through all this complexity in arriving at the perfection point mile, and that is by using the same technique that Péronnet and Thibault used for the marathon. While we don't feel it's legitimate for the marathon, it's reasonable to use it for the mile because it's a much shorter duration race and the equations apply throughout. Using their formulae for calculating power output for an ideal athlete varying over time during a race, the prediction for existing runners is the capability to run the mile in 3:41.96, which is not far off the present record of 3:43.13. By 2028 it should be down to 3:33.29 and by 2040, 3:29.84.

Extrapolating to some unknown time in the future but staying within the present stage of human evolution, the perfection point for the mile is 3:18.87.

"Blistering" barely begins to describe how fast that is for a mile runner. It's almost exactly the same speed that won the women's 100-meter *sprint* in 1925, and a weekend bicyclist probably wouldn't average it during a twenty-five-mile ride.

Will it ever really happen? It's doubtful, to a greater extent than for some of the other perfection points we discuss, because 3:18.87 is an "upper bound" rather than a prediction. What that means is that we can't say for certain whether a mile will ever actually be run this fast, but what we do know with absolute certainty is that none will ever be run that's faster.

Then again, if someone had told you just five years ago that a human would hold his breath for eleven minutes and thirty-five seconds, it would have sounded just as ridiculous.

ARE PERFECTION POINTS REACHABLE?

There isn't a record in existence that hasn't been broken.
—CHAY BLYTH

I knew my record would stand until somebody broke it.
—YOGI BERRA

We've had some fun trying to predict at what point and under what circumstances athletes might achieve absolute perfection in a variety of sports, using *absolute* in the scientific sense to mean "cannot get any better." This naturally raises the question of whether that can actually happen. The obvious objection is that, no matter how well an athlete performs, someone else can still conceivably do better.

There are some records that were once thought unbeatable. At the 1968 Summer Olympics in Mexico City, Bob Beamon broke the world long jump record with a stunning leap of 29 feet 2½ inches. To put that in perspective, consider that the record had been broken fifteen times since 1901, with an average increase of 1.9 inches. The biggest single increase was six inches.

Beamon broke the record by 21¾ inches.

A leap of that distance was so unimaginable that the optical measuring system wasn't set up to record it: It had to be measured manually. One journalist called Beamon "the man who saw lightning." Lynn Davies, the defending world champion, said to him, "You have destroyed this event," and Dick Schaap wrote a book about it, called *The Perfect Jump*. It was generally thought that Beamon's jump was a one-off anomaly that would never be equaled, much less bettered, especially in light of the fact that it had been done in the thin air of 7,349-foot-high Mexico City.

Twenty-three years later Mike Powell jumped two inches farther, and he did it in Tokyo, which is over 7,000 feet lower than Mexico City. Stories like that would seem to be a pretty good indicator that there are no safe records.

Then again, it's also a safe bet that there are physical limits to human athletic achievement, which is what we tried to demonstrate in this book. The argument against that point of view is that there is a mental component which, while elusive and unmeasurable, is nevertheless real and critically important. That argument is a good one and we use it in dropping the perfect 100-meter sprint to 8.99 seconds.

At the end of every major athletic event, sportscasters and analysts typically try to rationalize why the winner won. It's usually some variant of how much harder he trained than all the others or that his physical gifts are superior or that he was "hungrier."

The truth, however, is that they don't really know. Line up the top five finishers of the New York City Marathon or the Ironman

World Championships or the Tour de France or the 10,000-meter Olympic speed skate and what you'll discover is this: They all work hard, they're all physically gifted, and they're all hungry. None of those factors explains anything. If you had at your disposal a complete set of data on those athletes, showing such things as VO_2 max, muscle composition, aerobic efficiency, a compelling motivation to excel, etc., there is no way you could possibly figure out who was going to win. At the highest levels, they're all about the same in virtually every physical category, and even where there are differences, they don't correlate very highly with performance.

The difference between first and fifth place among elite athletes comes down to the mental side rather than the physical. The ability to stay relaxed under intense pressure is critical to peak performance. Self-confidence is equally important: Sports is littered with the remains of superbly conditioned athletes who were so unsure of themselves that they manufactured excuses for losing even before the competition began. Think about athletes who partied the night before a big event or got injured doing something abysmally and inexplicably stupid the week prior, and you're likely looking at somebody who was worried about losing instead of excited about winning. Champions always want the ball, and they're never afraid of looking bad.

Given the mysterious intangibility of the mental aspect of sports and the fact that it is at least as important as the physical, is there any point in trying to define the perfection point?

The answer is *Yes*, for the simple reason that the real role of the mental side is not to add speed or power but to ensure that the physical side is acting to its fullest potential. Athletes often call this "getting out of your own way." Obviously there's a lot more to it in skill sports, where the conscious mind is precisely guiding muscles that are making fine movements, but even in those cases, there are purely physical assumptions that define perfection. Take as an example a downhill skier negotiating a two-mile slope at speeds as high as 75 mph. The mind is fully involved as the competitor chooses a line, skis perilously close to the gates, decides how wide to take turns and how tightly to set his edges. But no matter how well he does, there's a physical limit to how fast he can go, because his speed is the result of gravity and there's nothing he can do to influence that. What the downhiller is actually trying to do is minimize friction by skiing smoothly, minimize air resistance by keeping a good tuck, and minimize the total distance traveled by assuming the straightest line possible.

The same is true in all the other sports we discussed. No matter how "mentally tough" a baseball player is, there's a limit to how fast it's possible to swing a bat and therefore a limit to how far he can hit a ball. "Mental toughness" gets him closer to that limit, but it can't put him past it.

So there's little doubt that perfection points exist and that they can be discerned. Whether we've arrived at them correctly remains to be seen, and we're hopeful that others will take up the challenge and try to refine them. That will happen over time and with much vigorous debate that will likely never end as more evidence is uncovered and better models are developed.

That leaves us with a bigger and much more intriguing question than whether perfection points exist, and that is: Will human beings ever actually achieve them?

The answer is *No.*

In order to see why, it's necessary to understand the concept of an *asymptote.* The term derives from a Greek word meaning "not meeting." Simply stated, an asymptote is a point you can get closer and closer to but never quite reach. You might remember an old children's riddle about a prince trying to get to the king's daughter. In his first step, he covers half the distance to her; in his second step, he covers half the remaining distance, and so on. Because he goes half the remaining distance with each step, he'll never reach her, but he gets closer and closer, by a smaller and smaller amount. This is how an asymptote works.

It's the same with world records in sports. Early in a sport's history, records get broken frequently and by wide margins. Over time, as athletes begin inching closer and closer to the limits of human potential, new records tend to occur over longer intervals and by smaller margins. Take a look at the progression in 100-meter sprint world records from 1912 through the Beijing Olympics of 2008:

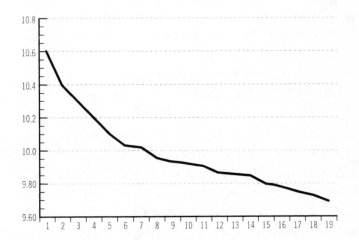

In the early days, there were huge drops. The first five records saw nearly 0.60 seconds of improvement. The last five records set were less than a fifth of that. The curve flattens out as improvements get smaller but, like the prince in our riddle, it keeps getting closer.*

---

* Usain Bolt's insane 9.58 in 2009 made a mockery of the whole curve, but I'm trying to make a point here. It *will* flatten out again.

The job we set for ourselves in this book was to find the asymptote for each sport covered. But it's not just a matter of coming up with a ridiculous number we know can never be reached, because that's not what an asymptote is. We could say that no one will ever run the 100-meter sprint in seven seconds, and we'll certainly be right, but we won't have shed any light on the subject, because the same is true of 7.5 seconds, and 8.0 seconds, and 8.2 seconds.

What we tried to do is to find the highest point possible that will never be breached but to which we will come tantalizingly close. It's the point at which absolute perfection is reached, where the tank is completely empty of gas, and beyond which no further improvement is possible.

Such a point exists, but if reaching it is impossible, why bother trying to find it?

Because it gives us something to shoot for.

Breaking a world record is always a stunning achievement in a mature sport, but we gauge existing records by what's gone on before. Doing so tells us that this athlete is the best right now, but it tells us nothing about where he—and the species—stands with respect to what's possible. Bolt's 9.69 at Beijing was incredible, but if you go back over the hours of commentary that surrounded it, you'll discover that most of the conversation was about what was still possible. Breaking 9.7 had been thought to be years away, but people started to rethink that and even dared speculate aloud as to when 9.60 would be breached and whether 9.50 *could* be. When Bolt then ran 9.58 the following year, the debate started up once again, and the question that was hanging in the air begging to be answered was, Where does it all end? When does the race run out of steam? We know Bolt is the best that ever was, because the numbers speak for themselves, but when it's no longer fun to measure him against other sprinters, the only thing left is to measure him against the best that could ever be.

Except that no one had any idea what that was. Now, they do: It's 8.99 seconds, as we showed in chapter 1. Records will continue to be broken in the "fastest man on earth" competition, but the conversation can now shift from "How much better was he than the last guy?" to "How close is he getting to the fastest run possible?"

Clearly we still have a ways to go. The same is true in every sport we talked about. But that brings us to another reason for trying to find perfection points: Defining them enhances the likelihood that we're going to get better faster.

Barriers are often more psychological than physical. When we thought a sub-four-minute mile was unlikely, it was hard for us to get there. Once we did, the records fell like hail. We just needed to know it was possible.

Now that Usain Bolt has shattered our preconceptions of what's possible in the 100 meters, those records will fall harder than they did before. And knowing how far we are from perfection will make them drop even faster. Records can be seen as stepping stones rather than ends in themselves, and they can become measures of how close we're creeping up on perfection. The good news for sports fans is that we will never get there, and we'll never stop breaking records. They might get stuck for a while—Jesse Owens's mark in the long jump stood for twenty-five years and Bob Beamon's for twenty-three—but eventually they're going to get broken.

Because the perfection point is not a measure of where we end . . . just where we're going.

ACKNOWLEDGMENTS

This book was not possible without the help and support of the following people:

Lee Gruenfeld, my coauthor, who was instrumental in writing, researching, and, most important, crafting this book. The countless hours Lee and I spent discussing, brainstorming, and challenging each other helped form a friendship that I will cherish for the rest of my life. He is truly one of the smartest people I have ever met, and I consider myself incredibly blessed to have met him and to have had the chance to work with him so closely.

Mickey Stern, my business partner and brother-in-law, who helped build Sport Science into a world-renowned brand. He and I have grown BASE Productions out of the basement of my parents' house and turned it into one of the most prolific production companies in Hollywood. I'll never be able to express how much respect I have for him and how grateful I am that we are partners, family, and friends.

Rose and John Brenkus, my parents, for their unconditional love and support.

Mary Stern, my sister, for always encouraging me.

Luke and Lexie Stern, my nephew and niece, for reminding me how grateful we need to be for our family!

Bryce and Arabella, my children, for always helping me do my best and for constantly keeping me laughing.

And, most important, Lizzie, my wife. I sat next to Lizzie on a plane seven years ago and fell madly in love. Turned out that we lived two blocks from each other on the same street in LA. Sometimes, God throws you a softball. She is the single most amazing person I've ever met, and I'm in awe of her every day.

I am deeply indebted to a number of outstanding experts whose wisdom, insight, and enthusiasm were invaluable to our understanding of the complexities of human athletic performance:

Dr. Bassil Aish, chief medical advisor for *Sport Science* on ESPN

Ron Allice, director of USC Track and Field

Tod Baden, owner/director of Synergy Sports in Toledo, Ohio

Cindy Bir, PhD, Wayne State University, for being there from the very beginning

David Morgan Brenner, Navigator Advisory Services, Encinitas, California

Morgan Brenner (David's father), author of *College Basketball's National Championships* and *The Majors of Golf: Complete Results of the Open, the U.S. Open, the PGA Championship and the Masters, 1860–2008*

Jeff Butt, president of the Canadian Powerlifting Union

Dr. Ben Domb, pioneering orthopedic surgeon at Hinsdale Orthopaedic Associates, and his associate, Dr. Adam Brooks

Bill Graham, U.S. record holder in static apnea, Kailua-Kona, Hawaii

Michael Gross and Nikolai von Keller, researchers to the stars

Jon Hall (www.powerliftingwatch.com)

Perry Husband, one of baseball's best hitting coaches

Sean Hutchison, CEO and head coach, King Aquatic Club in Seattle, 2009 U.S. World Championship Team head coach,

2008 U.S. Olympic coaching staff member and two-time USA Swimming George Haines Award recipient

Porter W. Johnson, professor of physics, emeritus, at the Illinois Institute of Technology

Dr. Jason Karp, San Diego State University cross-country coach, founder and host of the VO$_2$max Distance Running Clinic and the San Diego Personal Training Summit

Dan Lieberman, professor of human evolutionary biology, Harvard University

Russell Mark, USA Swimming's director of national team performance support

David McKinley, media editor, Fullerton College

Dr. Scott McLean, Southwestern University

Scott Mendelson, current world record holder in the raw (unassisted) bench press

David Sandler, president and cofounder of StrengthPro, Inc., chairman of the Arnold Strength Training Summit at the Arnold Classic, and former assistant strength and conditioning coach and head of baseball at the University of Miami

Angela Simons, USA Powerlifting Secretary and National Referee

Joseph Warpeha, assistant professor and director of the Exercise Physiology Laboratories at the College of St. Scholastica, and USA Powerlifting national referee

I also need to thank the Emmy Award–winning "Sport Science" team (in no particular order): Murray Oden, John Davis, Tim Gordon, Scott Carrithers, Mark Morris, Carlos Contreras, Mark Kadin, Rob Dorfman, Rod Park, Erick Geisler, Annie Tang, Randa Cardwell, Dan Schulman, Brian King, Brian Swanson, Scott Bramble, Tracey Broderick, Frank Openchowski,

Brent Dones, Tom Buderwitz, Matt Murphy, Elita Fielder, Janice Tucker, Brenda Arson, Teryn Hogland, Eric Flathers, Nick Titus, Nick Tramontano, Tom Heigl, Matt Hayes, Dave Navarro, Johnny Camilo, Joanna Probst, Christel Thompson, Jason Mergott, Brent Long, Steven Roessler, Zach Hollifield, Robert Penzel, Aaron Wigo, Corey Becker, Chip Mullaney, David Koenig, Rhett Blankenship, Tim Libeau, Renee Maguire, Elizabeth Ventura, Gary Nepa, and everyone who is part of the best production and post-production team in Hollywood.

From Endeavor Talent Agency: Shawn Coyne and Greg Horangic

From William Morris Endeavor: Lance Klein and Jay Mandel

From Loeb & Loeb: Scott Edel

From ESPN: John Skipper, Keith Klinkscales, John Walsh, Joan Lynch, Ron Wechsler, Lee Fitting

From Fox Sports Net: George Greenberg, Bob Thompson, and David Leepson

From HarperCollins: I'd like to thank Adam Korn for starting this project, and Matt Harper for brilliantly guiding it to completion.

CPSIA information can be obtained
at www.ICGtesting.com
Printed in the USA
LVOW08s1731131217
559454LV00007B/63/P